WORKSHEETS

FOR CLASSROOM OR LAB PRACTICE

CARRIE GREEN

INTERMEDIATE ALGEBRA

George Woodbury

College of the Sequoias

Boston San Francisco New York
London Toronto Sydney Tokyo Singapore Madrid
Mexico City Munich Paris Cape Town Hong Kong Montreal

Reproduced by Pearson Addison-Wesley from electronic files supplied by the author.

Publishing as Pearson Addison-Wesley, 75 Arlington Street, Boston, MA 02116.

ISBN-13: 978-0-321-52315-0
ISBN-10: 0-321-52315-6

1 2 3 4 5 6 OPM 11 10 09 08

Table of Contents

Name: Date:
Instructor: Section:

Chapter 1 REVIEW OF REAL NUMBERS

1.1 Integers, Opposites, and Absolute Value

Learning Objectives
1 Understand sets of numbers.
2 Determine the greater of two integers.
3 Find the absolute value of an integer.
4 Add and subtract integers.
5 Multiply and divide integers.
6 Simplify exponents.
7 Use the order of operations to simplify arithmetic expressions.

Key Terms
Use the most appropriate term or phrase from the given list to complete each statement in exercises 1-6.

real numbers	**natural numbers**	**whole numbers**	**greater than**
less than	**exponent**	**negative numbers**	**opposite**
integers	**absolute value**	**product**	**factors**
dividend	**divisor**	**quotient**	**base**

1. The ______________________ of a number is a number on the other side of 0 on the number line and is the same distance from 0 as that number.

2. The set of ______________________ is the set of natural numbers and 0.

3. A number a is ______________________ another number b if it is located to the right of b on the number line.

4. Numbers to the left of 0 on the number line are called ______________________.

5. In a division problem, the result is called the ____________________.

6. In exponential notation, the ____________________ tells us how many times the ____________________ is used as a factor.

Name: Date:
Instructor: Section:

Objective 1 Understand sets of numbers.
Objective 2 Determine the greater of two integers.

Write the appropriate symbol, either < or >, between the following integers.

1. -14 ___ -16 **1.**________________

2. 6 ___ -24 **2.**________________

3. -56 ___ -64 **3.**________________

4. -98 ___ 89 **4.**________________

Objective 3 Find the absolute value of an integer.

Find the absolute values.

5. $|-3049|$ **5.**________________

6. $|1207|$ **6.**________________

7. $|-4|$ **7.**________________

8. $|123|$ **8.**________________

9. $|-171|$ **9.**________________

Find the missing number, if possible. There may be more than one number that works, so find as many as possible. There may be no number that works.

10. $2 \cdot |?| = 8$ **10.**________________

Name: Date:
Instructor: Section:

11. $|?| - 5 = 15$

11.________________

Objective 4 Add and subtract integers.

Add.

12. $(-6)+(-3)$

12.________________

13. $(-4)+(-13)$

13.________________

14. $15+(-13)$

14.________________

Subtract.

15. $(-13)-(-15)$

15.________________

16. $(-32)-(-27)$

16.________________

17. $9-24$

17.________________

Simplify.

18. $15+(-6)-21-(-12)$

18.________________

Name: Date:
Instructor: Section:

19. $52-33-(-7)+(-54)$

19.______________

Objective 5 Multiply and divide integers.

Multiply.

20. $5 \cdot 24$

20.______________

21. $(-7)(-3)(-4)(-2)(-5)$

21.______________

22. $9(-1)(-3)(-5)(-7)$

22.______________

23. $-2(-13)$

23.______________

Divide.

24. $-264 \div (-11)$

24.______________

25. $805 \div (-7)$

25.______________

26. $-533 \div 41$

26.______________

Name: Date:
Instructor: Section:

27. $-1425 \div (-19)$ **27.**________________

Objective 6 Simplify exponents.

Simplify the given expression.

28. 10^6 **28.**________________

29. 10^{11} **29.**________________

30. $(-2)^8$ **30.**________________

31. -3^6 **31.**________________

32. $6^4 \cdot \left(\frac{1}{6}\right)^3$ **32.**________________

33. $\left(\frac{1}{10}\right)^{25} \cdot 10^{25}$ **33.**________________

Objective 7 Use the order of operations to simplify arithmetic expressions.

Simplify the given expression.

34. $\dfrac{6-2\cdot 9}{1-4^2}$ **34.**________________

Name: Date:
Instructor: Section:

35. $\dfrac{4(-2-9)+3\cdot 2}{4-5\cdot 2}$

35.________________

36. $-3[4(5-3)-7(9-2)]$

36.________________

37. $-[2(5-11)-(-29+8)]+20$

37.________________

38. $\dfrac{10\cdot 3-7\cdot 2}{2+6\cdot 7}$

38.________________

Find the missing number(s).

39. $?^{1967}=1$

39.________________

40. $?^2=\dfrac{121}{169}$

40.________________

Name: Date:
Instructor: Section:

Chapter 1 REVIEW OF REAL NUMBERS

1.2 Introduction to Algebra

Learning Objectives
1 Build variable expressions.
2 Evaluate variable expressions.
3 Understand and use the commutative and associative properties of real numbers.
4 Understand and use the distributive property.
5 Identify terms and their coefficients.
6 Simplify variable expressions.

Key Terms
Use the most appropriate term from the given list to complete each statement in exercises 1-4.

associative	**distributive**	**variable expression**	**commutative**
coefficient	**like terms**	**variable**	

1. For any real numbers x and y we know that $x + y = y + x$ by the ____________________ property.

2. A ____________________ is a combination of variables with numbers and/or arithmetic operations.

3. The ____________________ property allows us to group addition or multiplication in different ways.

4. The expressions $4x^2$ and $17x^2$ can be called ____________________ because they have the same variable factors.

Objective 1 Build variable expressions.

Build a variable expression for the following phrases. Let x represent the variable.

1. The quotient of 6 and a number **1.**________________

2. A number increased by 42 **2.**________________

3. Three times the difference of a number and 13 **3.**________________

Name: Date:
Instructor: Section:

4. The quotient of a number and three less than that number

4.________________

5. The admission charge to a fundraiser for a local charity is \$80/couple. If we let c represent the number of couples in attendance, build a variable expression for the amount of money taken in by the charity.

5.________________

6. B.J. McCay, a truck driver in the 1970's TV show *B.J. and the Bear*, charged \$1.50 per mile to haul cargo. If we let m represent the number of miles on a trip, build a variable expression for the charge to hire B.J. McCay to haul cargo.

6.________________

7. A tutorial service provides tutoring in mathematics to groups. For a one-hour session the service charges \$35 for the first person plus \$5 per each additional person. If we let n represent the number of students in a group, build a variable expression for the charge to tutor the group for a one-hour session.

7.________________

8. A taxi driver charges \$5 for the first two miles of a trip, plus \$0.80 per mile after that. If we let m represent the number of miles for a trip, build a variable expression for the charge for the trip. (You may assume in this case that m will be greater than 2 miles.)

8.________________

Objective 2 Evaluate variable expressions.

Evaluate the following algebraic expressions under the given conditions.

9. $3(x+h)^2-7(x+h)$ for $x=-3$ and $h=0.1$

9.________________

Name: Date:
Instructor: Section:

10. $\dfrac{(x+h)^2-5(x+h)+17}{h}$ for $x=-2$ and $h=0.001$

10.________________

11. $\dfrac{[(x+h)^2+5(x+h)-13]-[x^2+5x-13]}{h}$
for $x=8$ and $h=2$

11.________________

12. $(17z-15)-16z$ for $z=-343$

12.________________

13. $100(3x-2904)-300x$ for $x=8463$

13.________________

14. $x^2-5x+22$ for $x=7$

14.________________

15. m^2-36 for $m=-6$

15.________________

Name: Date:
Instructor: Section:

Objective 3 Understand and use the commutative and associative properties of real numbers.

Simplify, where possible.

16. $5a-2b+7a+3c-4b-2a+9c+16a$ **16.**____________

17. $2a-19a$ **17.**____________

18. $-4b+17-18b-36+9b$ **18.**____________

Objective 4 Understand and use the distributive property.

Simplify, where possible.

19. $3(4x+5)+6(7x+8)$ **19.**____________

20. $11(5+7x)-9(8x-13)$ **20.**____________

21. $5(4a-3b+7c+6)$ **21.**____________

22. $5(3v-4w+5x-11y+z)$ **22.**____________

Name: Date:
Instructor: Section:

Objective 5 Identify terms and their coefficients.

For the following expressions, simplify the expression, if possible, then
a) determine the number of terms;
b) write down each term; and
c) write down the coefficient for each.

23. $9x$

23a._______________

b._______________

c._______________

24. $x-2(x+1)+3(x+2)-4(x+3)+5(x+4)$

24a._______________

b._______________

c._______________

25. $3(5x-4y+8z+2)-2(x-6y+7z+3)$

25a._______________

b._______________

c._______________

26. $2x^4-3x^3+5x^2-8x+13$

26a._______________

b._______________

c._______________

Objective 6 Simplify variable expressions.

Simplify, where possible.

27. $3(y+4)$

27._______________

Name: Date:
Instructor: Section:

28. $12a-10-20a+34$ **28.**________________

29. $5c+17+16c+43$ **29.**________________

30. $-3(5a+5b-14c)+5b$ **30.**________________

Name: Date:
Instructor: Section:

Chapter 1 REVIEW OF REAL NUMBERS

1.3 Linear Equations and Absolute Value Equations

Learning Objectives	
1	Solve linear equations using the multiplication and addition properties of equality.
2	Solve linear equations using the five-step general strategy.
3	Identify linear equations that are identities or contradictions.
4	Solve a literal equation for a specified variable.
5	Solve absolute value equations.

Key Terms

Use the most appropriate term or phrase from the given list to complete each statement in exercises 1-5.

contradiction	**identity**	**negative**	**zero**	**one**
simplify	**positive**	**solve**	**null set**	**two**

1. The equation $|X| = a$ has ____________________ solution(s) if a is a negative number.

2. When the solutions set for an equation is the empty set, the equation is called a ____________________.

3. The solution set for a(n) ____________________ is the set of all real numbers.

4. To ____________________ an equation means to find its solution set.

5. The symbol $\varnothing$ represents the ____________________.

Objective 1 Solve linear equations using the multiplication and addition properties of equality.

Solve using the multiplication property of equality.

1. $-12x = 20$ **1.**________________

2. $15 = -3x$ **2.**________________

Name: Date:
Instructor: Section:

3. $-96 = -16t$ 3.________________

4. $-t = 45$ 4.________________

Solve using the addition property of equality.

5. $t + 11 = 3$ 5.________________

6. $y + 12 = -5$ 6.________________

7. $5 + x = 29$ 7.________________

8. $8 + n = 1$ 8.________________

Objective 2 Solve linear equations using the five-step general strategy.

Solve.

9. $4x + 45 = 14x + 15$ 9.________________

Name: Date:
Instructor: Section:

10. $\frac{2}{5}(3x-32)=\frac{1}{2}x+6$

10.________________

11. $\frac{2}{3}(12x-9)+\frac{3}{5}(10-15x)=\frac{1}{2}(6x+8)$

11.________________

12. $\frac{1}{4}(3x+19)=\frac{3}{5}x+\frac{11}{5}$

12.________________

13. $4-5a=a-13$

13.________________

14. $-8y+8=37y-7$

14.________________

15. $\frac{1}{3}x+\frac{3}{4}=\frac{5}{24}x+\frac{3}{2}$

15.________________

Name: Date:
Instructor: Section:

16. $2x-\frac{7}{6}=\frac{5}{4}x+\frac{67}{12}$

16.________________

Objective 3 Identify linear equations that are identities or contradictions.

Solve.

17. $(5x+2)-(8x-2)=-2(x+2)-x$

17.________________

18. $(3x+2)+(4x+7)=2(x+5)+(5x-1)$

18.________________

19. $51x-7-3x=17+48x-24$

19.________________

20. $5y-8=37+5y$

20.________________

Objective 4 Solve a literal equation for a specified variable.

Solve the following literal equations for the specified variable.

21. $\frac{1}{2}x+\frac{1}{3}y=\frac{2}{3}$ for y

21.________________

Name: Date:
Instructor: Section:

22. $\frac{3}{5}x - \frac{1}{4}y = \frac{5}{2}$ for y

22.________________

23. $P = 4s$ for s

23.________________

24. $A = L \cdot W$ for L

24.________________

25. $S = 2\pi r^2 + 2\pi rh$ for h

25.________________

26. $I = Prt$ for P

26.________________

Objective 5 Solve absolute value equations.

Solve.

27. $|x+5| = 9$

27.________________

28. $|x+3| - 5 = 7$

28.________________

29. $|3x+17| = 0$

29.________________

Name: Date:
Instructor: Section:

30. $|5x-9|=-15$ **30.**________________

31. $|2x-15|-8=-3$ **31.**________________

32. $|2x-9|-24=-17$ **32.**________________

33. $|m+5|+6=27$ **33.**________________

34. $|3a+8|=|4a-4|$ **34.**________________

35. $|2a+7|=|3a \quad 5|$ **35.**________________

36. Find an absolute value equation whose solution is $\left\{-\frac{13}{2}, \frac{3}{2}\right\}$. **36.**________________

37. Find an absolute value equation whose solution is $\{7\}$. **37.**________________

Name: Date:
Instructor: Section:

Chapter 1 REVIEW OF REAL NUMBERS

1.4 Problem Solving: Applications of Linear Equations

Learning Objectives
1 Understand the six steps for solving applied problems.
2 Solve problems involving unknown numbers.
3 Solve problems involving geometric formulas.
4 Solve problems involving consecutive integers.
5 Solve problems involving motion.
6 Solve other applied problems.

Key Terms
Use the most appropriate term or phrase from the given list to complete each statement in exercises 1-3.

circumference **equilateral** **isosceles** **area** **consecutive**

1. A(n) ________________ triangle has three equal sides.

2. ________________ integers are 1 unit apart from each other on the number line.

3. The ________________ of a circle is the distance around the circle.

Objective 1 Understand the six steps for solving applied problems.
Objective 2 Solve problems involving unknown numbers.

Solve.

1. One number is 4 less than 5 times another number. If the sum of the two numbers is 68, find the two numbers.

1.________________

2. One number is 8 more than another number. If 3 times the smaller number is subtracted from 5 times the larger number, the difference is 84. Find the two numbers.

2.________________

Name: Date:
Instructor: Section:

Objective 3 Solve problems involving geometric formulas.

Solve.

3. An equilateral triangle has a perimeter of 45 cm. Find the length of each side of this triangle. **3.**________________

4. The third side of an isosceles triangle is 7 meters longer than each of the other two sides. If the perimeter of this triangle is 46 meters, find the lengths of the 3 sides. **4.**________________

5. Angles A and B are supplementary angles. Angle A measures 40° more than angle B. Find the measures of the two angles. **5.**________________

6. Angles A and B are supplementary angles. If angle A is 8 times angle B, find the measures of the two angles. **6.**________________

7. In addition to having three equal sides, an equilateral triangle has three angles that have the same measure. Find the measure of each angle in an equilateral triangle. **7.**________________

Name: Date:
Instructor: Section:

8. The smallest angle in a triangle is one-third as large as the largest angle. The other angle in the triangle is twice as large as the smallest angle. Find the measures of the three angles.

8.________________

9. If Tina was making a rectangular quilt out of 36 square blocks that each had a side of 6 inches, what arrangement would have the smallest perimeter? What is the perimeter?

9.________________

Objective 4 Solve problems involving consecutive integers.

Solve.

10. The sum of three consecutive even integers is 654. Find the three even integers.

10.________________

11. The sum of three consecutive integers is 165. Find the three integers.

11.________________

12. The sum of three consecutive odd integers is 87. Find the three odd integers.

12.________________

Name: Date:
Instructor: Section:

13. If the smallest of three consecutive odd integers is 75 less than twice the sum of the two larger integers, find the three integers.

13.________________

14. If the sum of five consecutive integers is 465, how many of the five integers are odd?

14.________________

15. If the sum of four consecutive integers is 454, how many of the four integers are odd?

15.________________

Objective 5 Solve problems involving motion.

For exercises 16-18, refer to the California Mileage Chart, which shows the distance between particular California cities in miles.

Miles	San Francisco	San Diego	Sacramento	Palm Springs	Needles	Los Angeles	Fresno
Barstow	420	175	415	125	145	115	245
Fresno	190	340	170	325	385	220	
Los Angeles	385	125	385	110	260		
Needles	560	320	55	210			
Palm Springs	485	140	495				
Sacramento	85	505					
San Diego	505						

To use this chart to find the distance between two cities, find a row in the table that contains one of the cities and a column that contains the other city. The intersection of this row and column is the distance between the two cities. For example, the distance between Los Angeles and Needles is 260 miles.

Name: Date:
Instructor: Section:

16. If it took Ross 4 hours to drive from Los Angeles to Fresno, what was his average speed for the trip?

16.________________

17. Linda drove from Palm Springs to Fresno, and then continued on to Sacramento. If she drove at an average speed of 55 miles per hour, how long did the trip take her?

17.________________

18. Cassandra drove from her home in Palm Springs to San Diego in 2 hours. If she increased her speed by 10 miles per hour on the trip home, how much quicker was the return trip?

18.________________

Objective 6 Solve other applied problems.

Solve.

19. In the last year Janet's stock portfolio has increased in value by $1842. If the value of her stock portfolio is now $23,056, what was the value of her stock portfolio one year ago?

19.________________

20. Juan has a change jar on his dresser and it contains pennies, nickels, dimes and quarters. The number of nickels is 1 more than twice the number of pennies. The jar has 2 more dimes than pennies. The number of quarters is 4 less than 3 times the number of pennies. If the value of all of the coins is $11.73, how many pennies, nickels, dimes and quarters are there in the jar?

20.________________

Name: Date:
Instructor: Section:

21. Kenny and Joyce are at a casino. Joyce has 7 more $5 chips than Kenny, and together they have $125 in $5 chips. How many $5 chips does Joyce have?

21._______________

Name: Date:
Instructor: Section:

Chapter 1 REVIEW OF REAL NUMBERS

1.5 Linear Inequalities and Absolute Value Inequalities

Learning Objectives

1. Graph the solutions of a linear inequality on a number line and express the solutions using interval notation.
2. Solve linear inequalities.
3. Solve compound linear inequalities.
4. Solve applied problems involving inequalities.
5. Solve absolute value inequalities.

Key Terms

Use the most appropriate term or phrase from the given list to complete each statement in exercises 1-3.

open **closed** **absolute value** **linear inequality** **compound**

1. When solving an absolute value inequality, isolate the __________________ before rewriting the inequality as a compound inequality.

2. Use a(n) __________________ circle to represent that the endpoint of an interval is included in the graph of the solution of an inequality.

3. A__________________ inequality is made up of two simple inequalities.

Objective 1 Graph the solutions of a linear inequality on a number line and express the solutions using interval notation.

Graph each inequality on a number line, and write the inequality in interval notation.

1. $x < 2$ **1.**__________________

2. $x \geq -4$ **2.**__________________

Name: Date:
Instructor: Section:

Objective 2 Solve linear inequalities.

Solve each inequality. Graph your solution on a number line, and write your solution in interval notation.

3. $x-1\leq -4$

3._______________

4. $x-3<6$

4._______________

5. $x+7>4$

5._______________

6. $x+6\geq 4$

6._______________

7. $-5x\leq 25$

7._______________

8. $3x+3<21$

8._______________

9. $3x+4>-7$

9._______________

10. $3x-4(2x+7)>3(2x-3)-8$

10._______________

Name: Date:
Instructor: Section:

Objective 3 Solve compound linear inequalities.

Solve each inequality. Graph your solution on a number line, and write your solution in interval notation.

11. $-8 < x - 7 < 8$

11.__________

12. $-7 < x + 1 \le 8$

12.__________

13. $10 < 3x + 16 \le 19$

13.__________

14. $-11 < 8 - x < 7$

14.__________

15. $4 \le 5 - 2x \le 11$

15.__________

16. $-10 < 5 - 3x < 23$

16.__________

17. $4x + 13 \le -7$ or $4x + 13 \ge 18$

17.__________

18. $-2x + 6 > 24$ or $8x + 33 > 17$

18.__________

Name: Date:
Instructor: Section:

Objective 4 Solve applied problems involving inequalities.

Solve.

19. A dance club has a rule that a person must be at least 18 years old to enter the club. Set up an inequality that shows the ages of people who are able to get in the club.

19.________________

20. A rental car company has two rental options for people who want to rent a car for 1 day. The first option is a flat fee of $49 with unlimited miles. The second option is $29 plus $0.10 per mile. Under what conditions is the first option ($49) less expensive?

20.________________

21. A local school district sets a goal that at least 90% of all incoming high school freshmen will graduate within four years. If there are 1264 incoming high school freshmen in the district this year, how many must graduate within four years in order for the district to reach its goal?

21.________________

Name: Date:
Instructor: Section:

22. The manager of a fast-food restaurant tells her employees that if they as a group generate an average of fewer than 2 complaints per day, then she will throw an ice cream party for them. During the first six days of the week there were 0, 1, 3, 2, 0 and 0 complaints. How many complaints can they receive on the seventh day and still qualify for the ice cream party?

22.________________

Objective 5 Solve absolute value inequalities.

Solve each inequality. Graph your solution on a number line, and write your solution in interval notation.

23. $|x| > 4$

23.________________

24. $|x+14| \leq 5$

24.________________

25. $|4x-7| > 14$

25.________________

26. $|4x-9| \geq 14$

26.________________

Name: Date:
Instructor: Section:

27. $|7x-2|+3 \geq 4$

27.________________

28. $|9-8x| \geq 14$

28.________________

29. $|4-x|+8 \leq 20$

29.________________

30. $|x+2| < \frac{5}{2}$

30.________________

Name: Date:
Instructor: Section:

Chapter 2 GRAPHING LINEAR EQUATIONS

2.1 The Rectangular Coordinate System; Equations in Two Variables

Learning Objectives
1. Determine if an ordered pair is a solution to an equation in two variables.
2. Plot ordered pairs on a rectangular coordinate plane.
3. Graph linear equations in two variables.
4. Find the x- and y-intercepts of a line from its equation.
5. Graph linear equations using their intercepts.
6. Graph horizontal and vertical lines.
7. Interpret the graph of an applied linear equation.

Key Terms

Use the most appropriate term or phrase from the given list to complete each statement in exercises 1-6.

ordered pairs	**Cartesian**	***x*-axis**	**standard**	**linear**
coordinates	**Descartes**	***y*-axis**	**origin**	**quadrants**
***x*-intercept**	***y*-intercept**	***x***	***y***	

1. The rectangular coordinate plane is also called the ____________________ plane.

2. On a rectangular coordinate plane, the ____________________ is horizontal.

3. The rectangular coordinate plane is divided into four ____________________.

4. The ____________________ form of a linear equation in two variables is $Ax + By = C$, where A, B, and C are real numbers and A and B are not both 0.

5. To find an x-intercept, substitute 0 for ____________________ and solve for ____________________.

6. A graph crosses the x-axis at a(n) ____________________.

Objective 1 Determine if an ordered pair is a solution to an equation in two variables.

Is the ordered pair a solution of the given equation?

1. $(2, 1)$, $y = 4x - 9$ **1.**____________

Name: Date:
Instructor: Section:

2. $(0,-2)$, $4x-5y=10$ **2.**____________

3. $(-5,1)$, $y=3x+16$ **3.**____________

4. $(4,\ -3)$, $-x+7y=17$ **4.**____________

Objective 2 Plot ordered pairs on a rectangular coordinate plane.

Plot the points on a rectangular coordinate plane.

5. $\left(\frac{1}{2},6\right), \left(-\frac{10}{3},6\right), \left(\frac{5}{9},-7\right), \left(-\frac{25}{4},\frac{23}{7}\right)$ **5.**

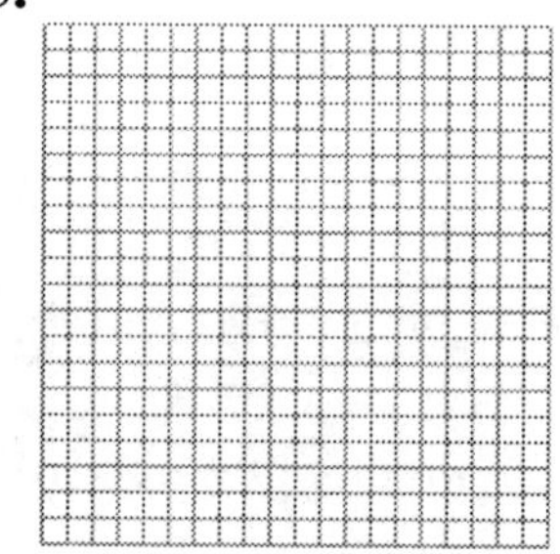

6. $(200,100)$, $(-300,200)$, $(150,-100)$, $(350,-425)$ **6.**

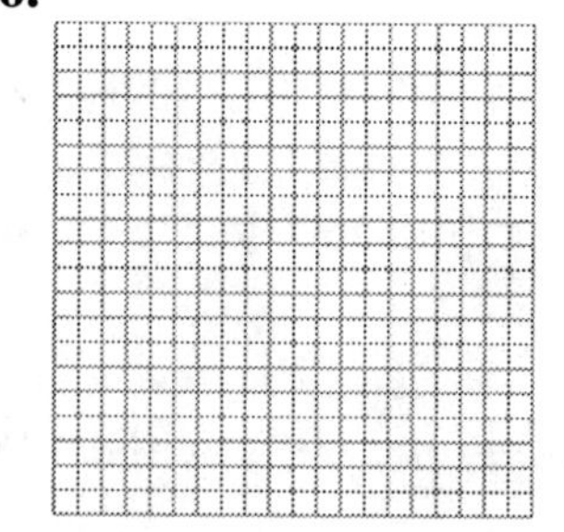

Name: Date:
Instructor: Section:

7. $(5,\ -3), (-2,\ 9), (-4,\ -6), (0,\ 4)$

7.

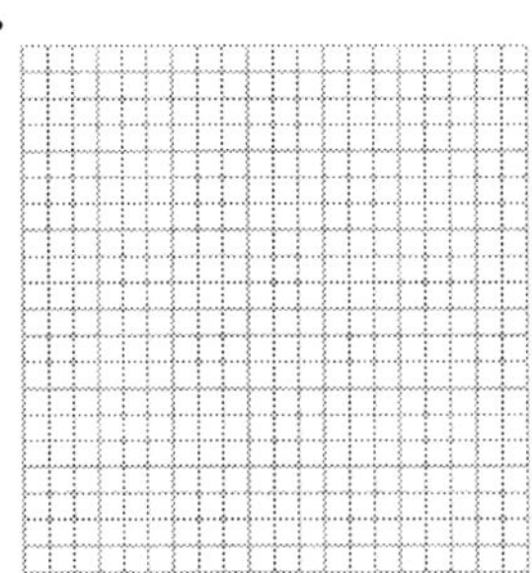

8. $\left(2,\ \frac{3}{2}\right), (-7,\ 0), (-5,\ 6), (6,\ -1)$

8.

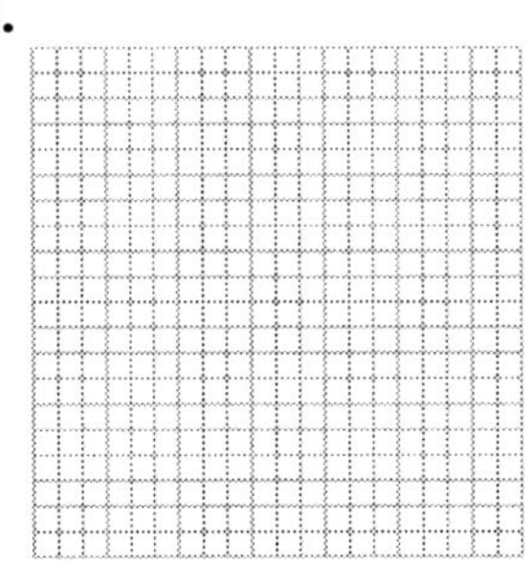

Find the coordinates of the labeled points A, B, C, and D.

9.

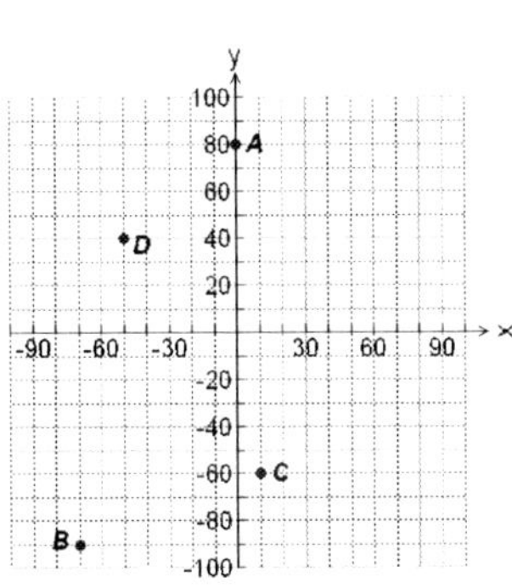

9.________________

____

____

____

10.

10.________________

____

____

____

Name: Date:
Instructor: Section:

Objective 3 Graph linear equations in two variables.

Using the given coordinate, find the other coordinate that makes the ordered pair a solution of the equation.

11. $\frac{3}{4}x+\frac{2}{3}y=1, (-4, y)$

11.________________

12. $\frac{2}{5}x-\frac{4}{9}y=14, (x, -18)$

12.________________

13. $5x-2y=2, (x, 4)$

13.________________

Complete the table with ordered pairs that are solutions.

14. $x+4y=9$

14.

x	y
-3	
0	
6	

Name: Date:
Instructor: Section:

Objective 4 Find the x- and y-intercepts of a line from its equation.

Find the x- and y-intercepts from the graph.

15. **15.**________________

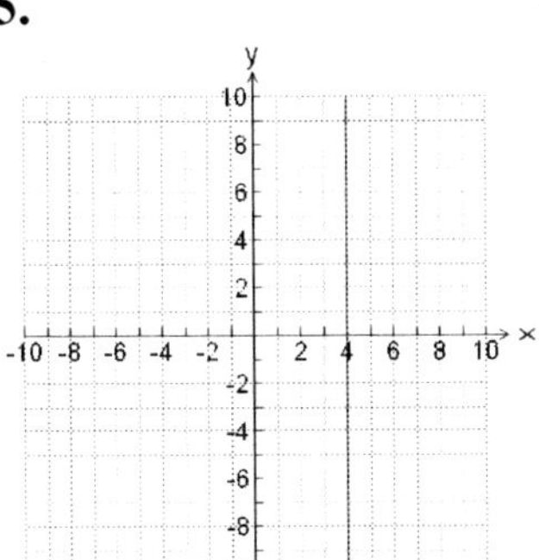

16. **16.**________________

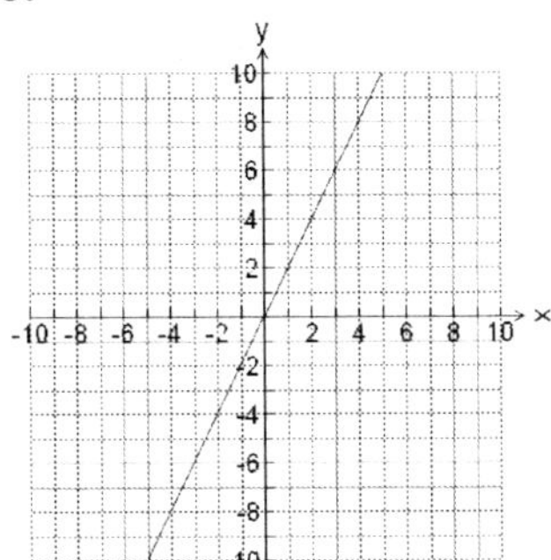

Find the x- and y-intercepts.

17. $y = -3$ **17.**________________

18. $x = 5$ **18.**________________

19. $3x - 7y = 0$ **19.**________________

20. $\frac{1}{12}x - \frac{1}{6}y = \frac{1}{4}$ **20.**________________

Name: Date:
Instructor: Section:

21. $x + y = 17$ **21.**________________

Objective 5 Graph linear equations using their intercepts.

Find the intercepts, and then graph the line.

22. $10x + 6y = -15$ **22.**________________

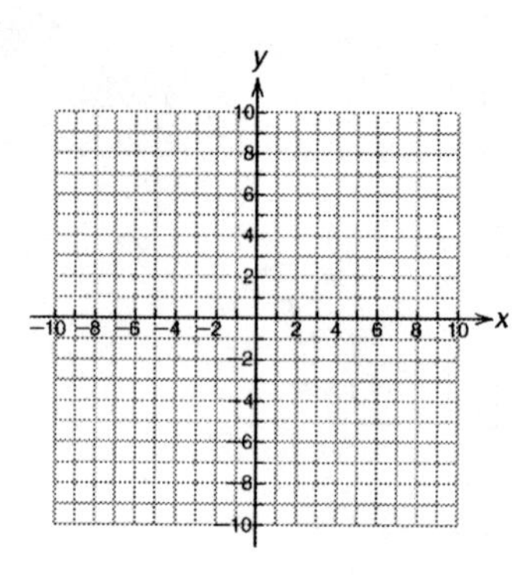

23. $14x + 10y = 35$ **23.**________________

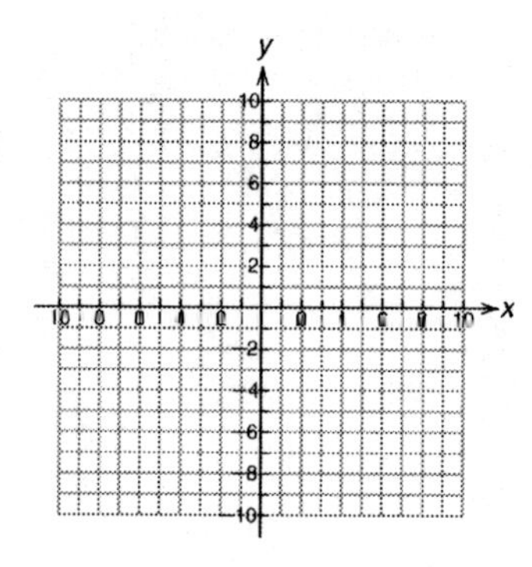

24. $7x - 3y = 21$ **24.**________________

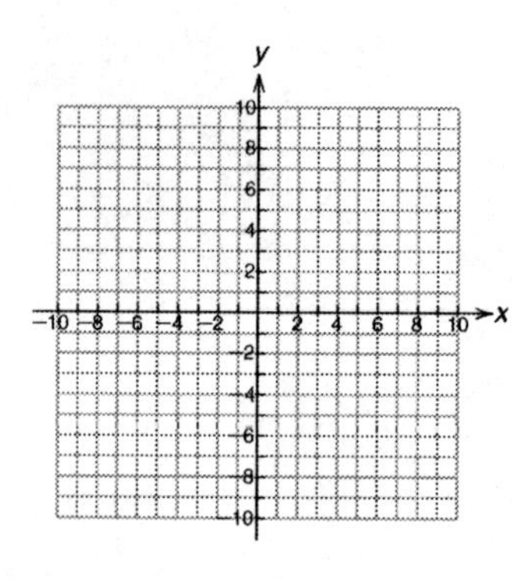

Name: Date:
Instructor: Section:

25. $2x+5y=10$

25.________________

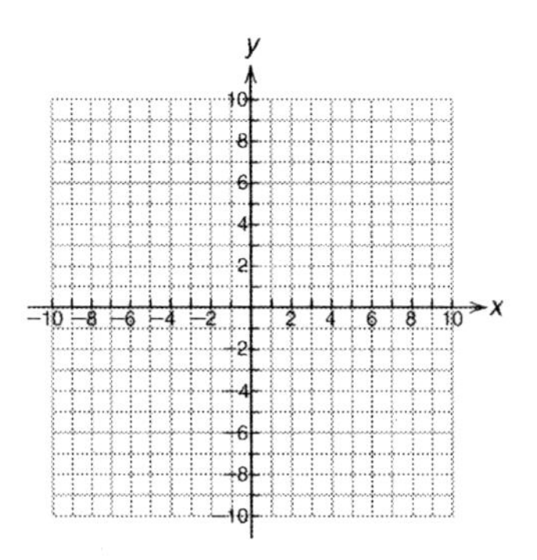

26. $x+2y=4$

26.________________

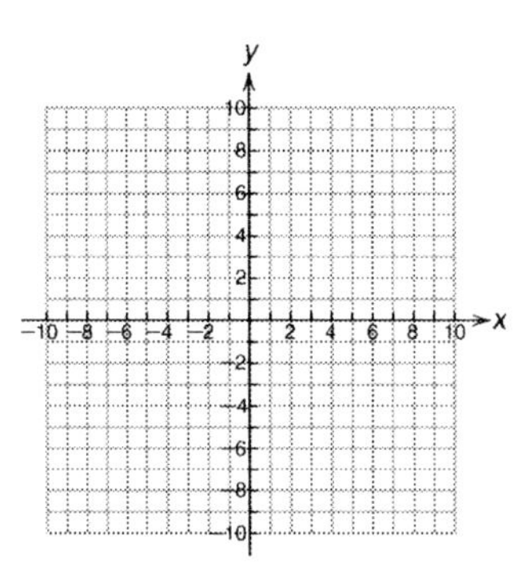

Find the intercepts, and then graph the line.

27. $y=-8x$

27.________________

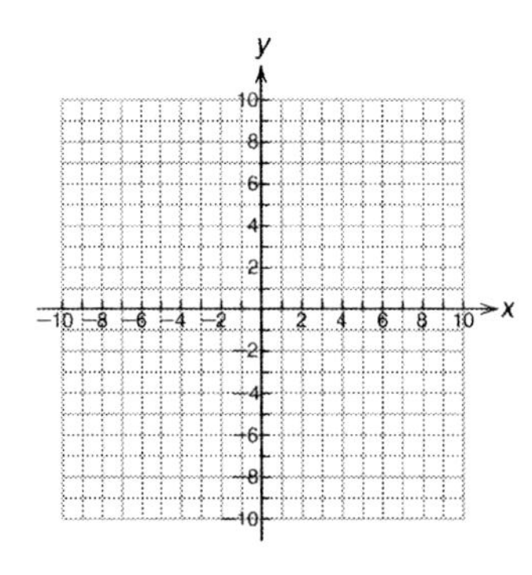

Name: Date:
Instructor: Section:

28. $y = 4x$

28.________________

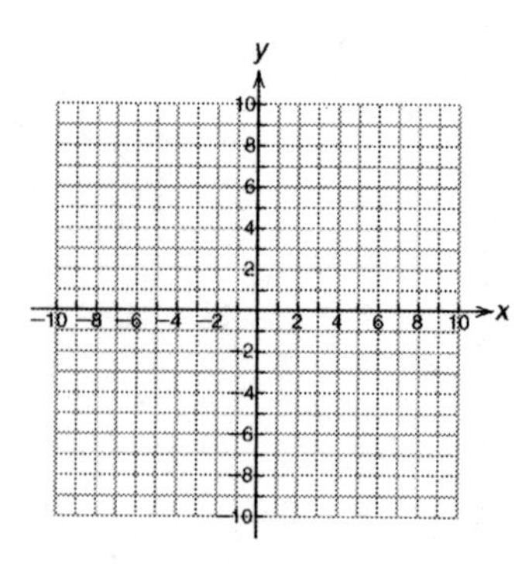

Objective 6 Graph horizontal and vertical lines.

Find the intercepts, and then graph the line.

29. $y = -7$

29.________________

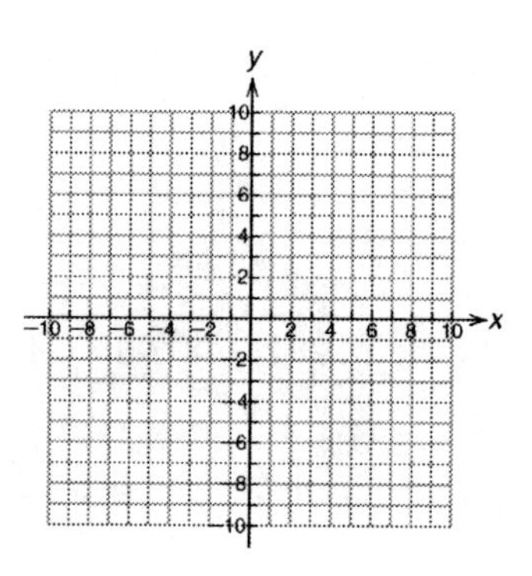

30. $y = \frac{13}{4}$

30.________________

Name: Date:
Instructor: Section:

Find the intercepts, and then graph the line.

31. $x = 4$

31. 

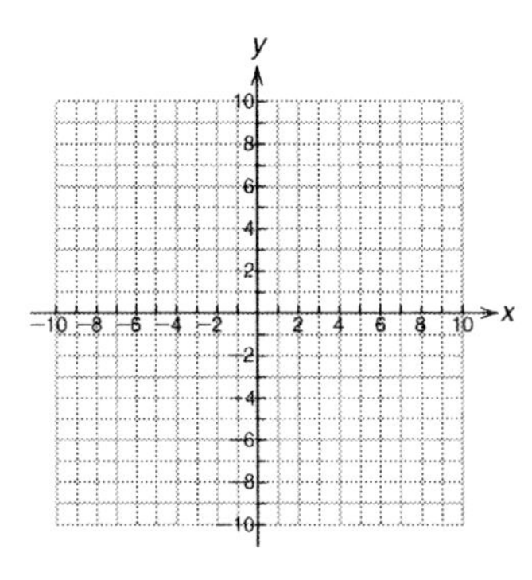

32. $x = 7$

32. ________________

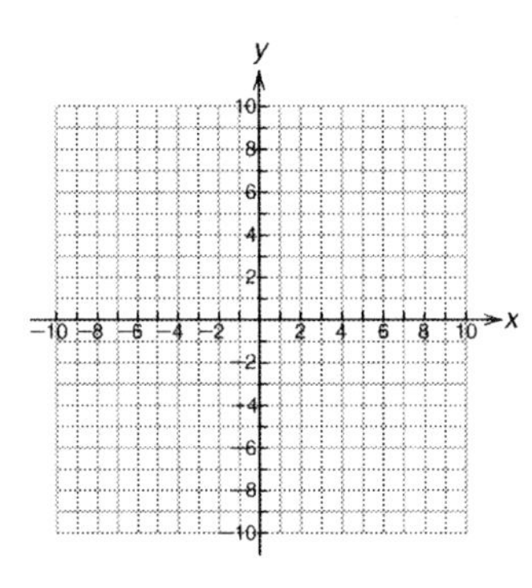

Objective 7 Interpret the graph of an applied linear equation.

Solve.

33. A long distance phone carrier charges \$0.03 per minute plus a connection fee of \$0.39 for long distance calls. The cost, y, of a long distance call lasting x minutes is given by the equation $y = 0.03x + 0.39$.

a. Find the y-intercept of this equation. Explain what this intercept signifies.

b. Find the x-intercept of this equation. Explain what this intercept signifies.

33a. ________________

____

____

b. ________________

____

____

Name: Date:
Instructor: Section:

Chapter 2 GRAPHING LINEAR EQUATIONS

2.2 Slope of a Line

Learning Objectives
1. Understand the slope of a line.
2. Find the slope of a line passing through two points using the slope formula.
3. Find the slopes of horizontal and vertical lines.
4. Find the slope and y-intercept of a line from its equation.
5. Graph a line using its slope and y-intercept.
6. Determine whether two lines are parallel, perpendicular, or neither.
7. Interpret the slope and y-intercept in real-world applications.

Key Terms
Use the most appropriate term or phrase from the given list to complete each statement in exercises 1-4.

negative reciprocals	**positive**	**undefined**	**negative**	**first**	**no**
perpendicular	**parallel**	**inverses**	**second**	**zero**	

1. The numerator of the slope formula can be thought of as the ________________ y-coordinate minus the ________________ y-coordinate.

2. A vertical line has ________________ slope.

3. Two nonvertical lines are ________________ if their slopes are equal.

4. The slopes $-\frac{1}{2}$ and 2 are ________________.

Objective 1 Understand the slope of a line.

Determine whether the given line has a positive or negative slope.

1. **1.**________________

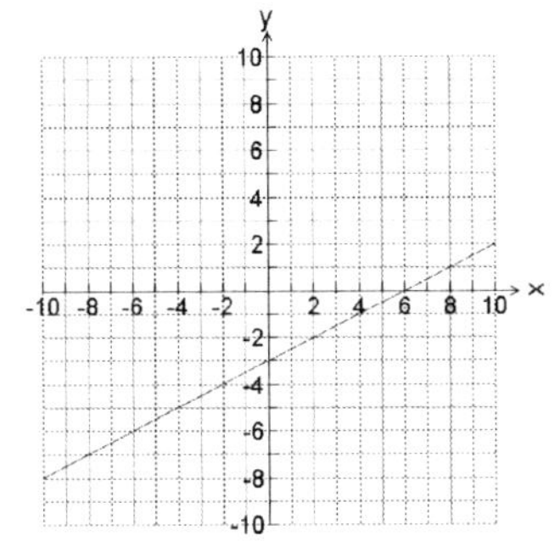

Name: Date:
Instructor: Section:

2.

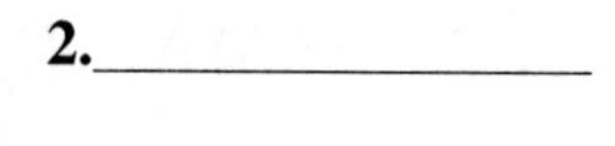

2.________________

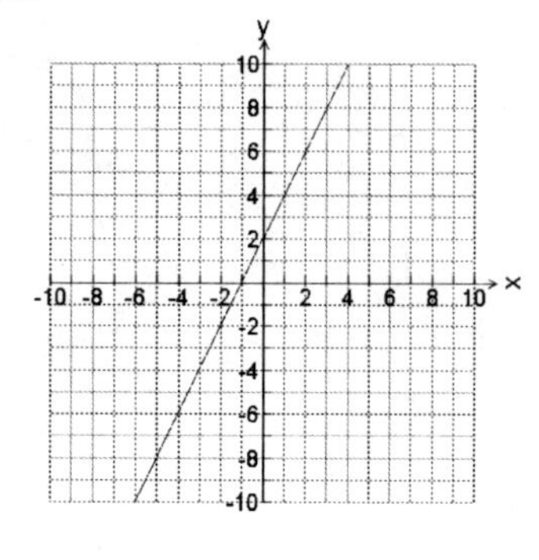

Find the slope of the given line. If the slope is undefined, state this.

3.

3.________________

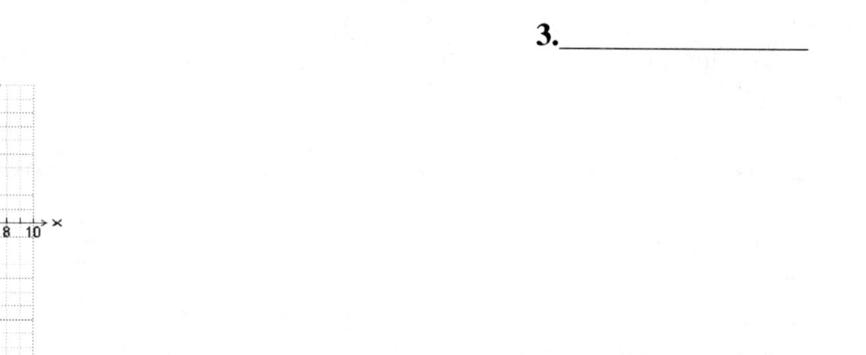

4.

4.________________

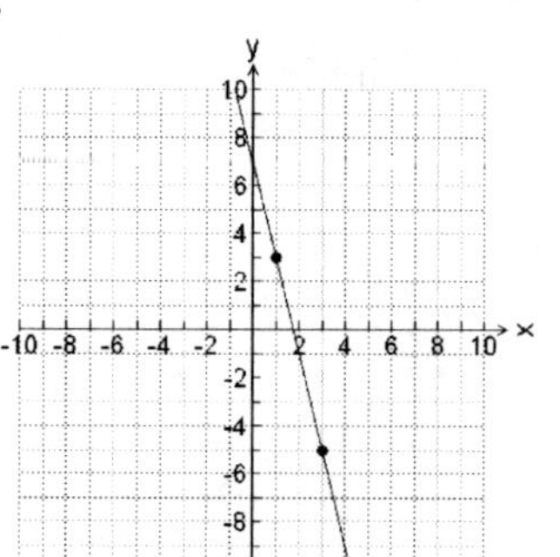

Objective 2 Find the slope of a line passing through two points using the slope formula.

Find the slope of a line that passes through the given two points. If the slope is undefined, state this.

5. $(1,3)$ and $(6,1)$

5.________________

Name: Date:
Instructor: Section:

6. $(3,0)$ and $(9,2)$

6.________________

7. $(-4,2)$ and $(1,\ 7)$

7.________________

8. $(-2,1)$ and $(4,4)$

8.________________

9. $(4,\ 5)$ and $(4,\ -5)$

9.________________

10. $(0,\ 0)$ and $(-2,\ 9)$

10.________________

11. $(-3,\ -7)$ and $(5,\ -5)$

11.________________

Objective 3 Find the slopes of horizontal and vertical lines.

Determine the slope of the line. If the slope is undefined, state this.

12. $y=-7$

12.________________

Name: Date:
Instructor: Section:

13. $x = 8$ **13.**________________

Objective 4 Find the slope and *y*-intercept of a line from its equation.

Find the slope and the y-intercept of the given line.

14. $y = 2x + 10$ **14.**________________

15. $y = 9x + 5$ **15.**________________

16. $4x - 3y = 9$ **16.**________________

17. $6x - 5y = 15$ **17.**________________

Find the equation of a line with the given slope and y-intercept.

18. Slope 7, *y*-intercept $(0,\ 1)$ **18.**________________

19. Slope 0, *y*-intercept $(0,\ 3)$ **19.**________________

Name: Date:
Instructor: Section:

Objective 5 Graph a line using its slope and y-intercept.

Graph using the slope and y-intercept.

20. $y = x + 4$

20.

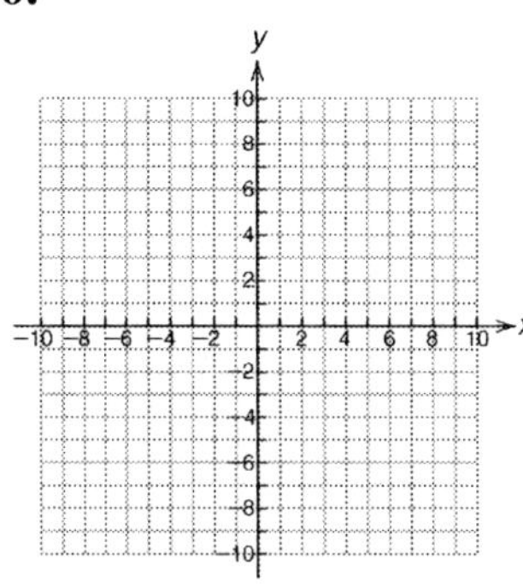

21. $y = -x + 5$

21.

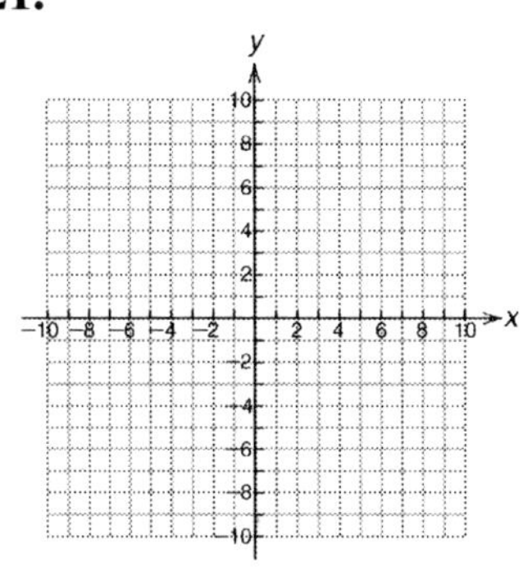

22. $y = -4x - 4$

22.

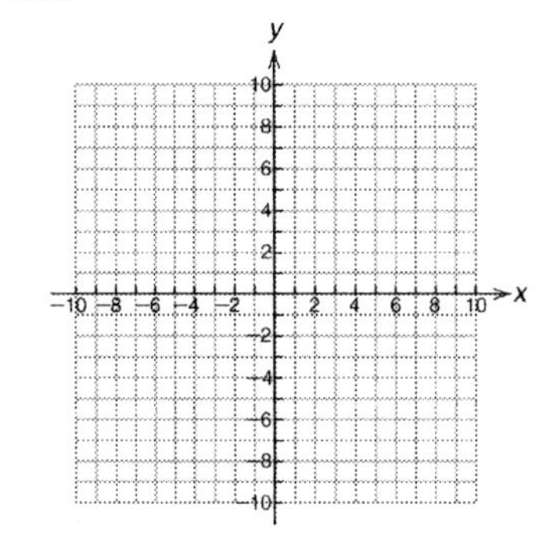

Name: Date:
Instructor: Section:

23. $y=-\frac{4}{7}x-3$

23.

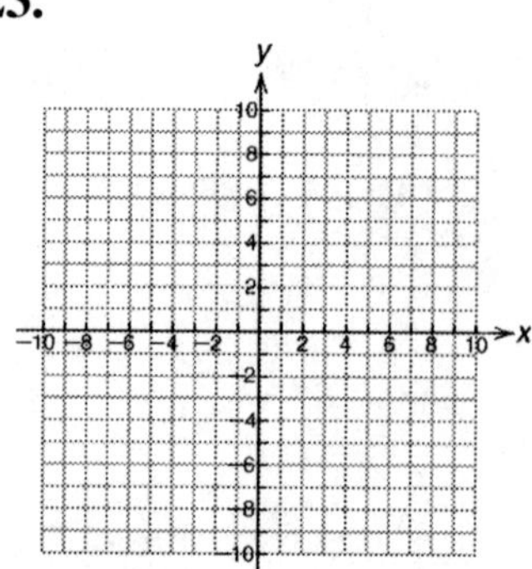

24. $y=x-3$

24.

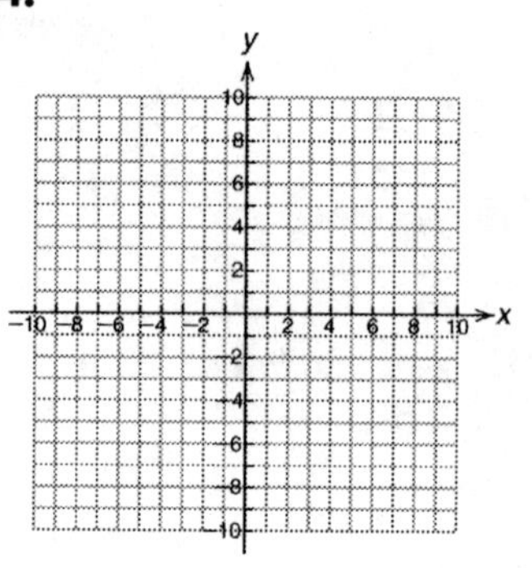

Objective 6 Determine whether two lines are parallel, perpendicular, or neither.

Are the two given lines parallel?

25. $y=2x+4,\ y=2x-5$

25.________________

26. $y=7,\ x=8$

26.________________

27. $y=-5x+1,\ y=5x+1$

27.________________

28. $2x+3y=5,\ 4x+6y=8$

28.________________

Name: Date:
Instructor: Section:

Are the two given lines perpendicular?

29. $y=\frac{1}{3}x+1$, $y=3x-1$ **29.**________________

30. $y=-\frac{2}{5}x+3$, $y=\frac{5}{2}x+4$ **30.**________________

31. $y=1$, $x=-2$ **31.**________________

32. $8x-y=7$, $x-8y=4$ **32.**________________

Are the two given lines parallel, perpendicular, or neither?

33. $y=3x+2$, $y=-3x-5$ **33.**________________

34. $y=7x-4$, $y=7x+3$ **34.**________________

35. $2x-4y=6$, $-\frac{1}{2}y=x+1$ **35.**________________

35. $5x+y=4$, $-5x+y=3$ **36.**________________

Name: Date:
Instructor: Section:

Objective 7 Interpret the slope and y-intercept in real-world applications.

Solve.

37. A ski slope drops 30 feet for every horizontal 50 feet. Find the slope.

37.________________

Name: Date:
Instructor: Section:

Chapter 2 GRAPHING LINEAR EQUATIONS

2.3 Equations of Lines

Learning Objectives	
1	Find the equation of a line using the point-slope form.
2	Find the equation of a line given two points on the line.
3	Find a linear equation to describe real data.
4	Find the equation of a parallel or perpendicular line.

Key Terms

Use the most appropriate term or phrase from the given list to complete each statement in exercises 1-2.

point-slope **standard** **slope** ***y*-intercept** ***x*-intercept**

1. The equation $y-3=7(x+1)$ is in ________________ form.

2. To find the equation of a line given two points on the line, first find the ________________ of the line.

Objective 1 Find the equation of a line using the point-slope form.

Write the following equations in slope-intercept form.

1. $5x+3y=-7$ **1.**________________

2. $-2x+4y=12$ **2.**________________

Find the equation of a line with the given slope and y-intercept.

3. Slope 7, y-intercept $(0,\ 5)$ **3.**________________

4. Slope -3, y-intercept $(0,\ -1)$ **4.**________________

Name: Date:
Instructor: Section:

Find the equation of a line with the given slope that passes through the given point. Write the equation in slope-intercept form.

5. Slope 2, through $(-2,-3)$ **5.**________________

6. Slope 4, through $(-3,-4)$ **6.**________________

7. Slope -1, through $(-6,1)$ **7.**________________

8. Slope 3, through $(-4,-4)$ **8.**________________

9. Slope 1, through $(-8,1)$ **9.**________________

10. Slope 0, through $(-5,6)$ **10.**________________

Objective 2 Find the equation of a line given two points on the line.

Find the slope-intercept form of the equation of a line that passes through the given points. Graph the line.

11. $(-8,14)$, $(12,-1)$ **11.**

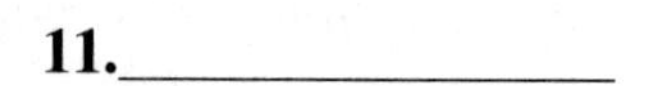

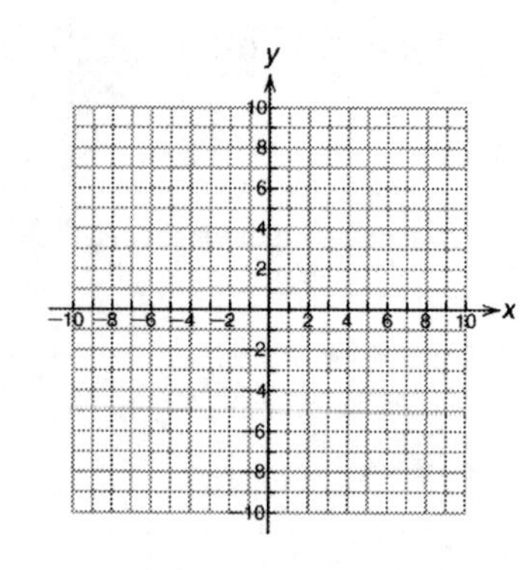

Name: Date:
Instructor: Section:

12. $(6,-2)$, $(-3,10)$

12.________________

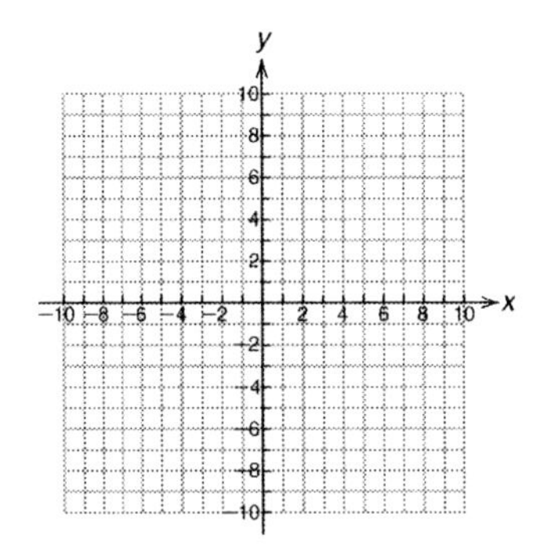

13. $(-1,-2)$, $(9,2)$

13.________________

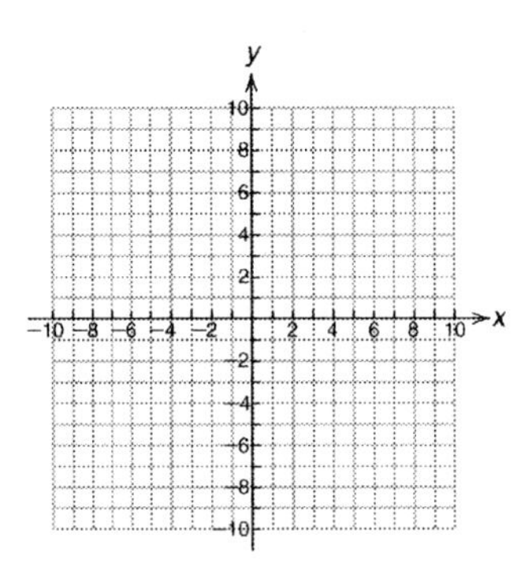

14. $(-6,\ -6)$, $(-2,\ -3)$

14.________________

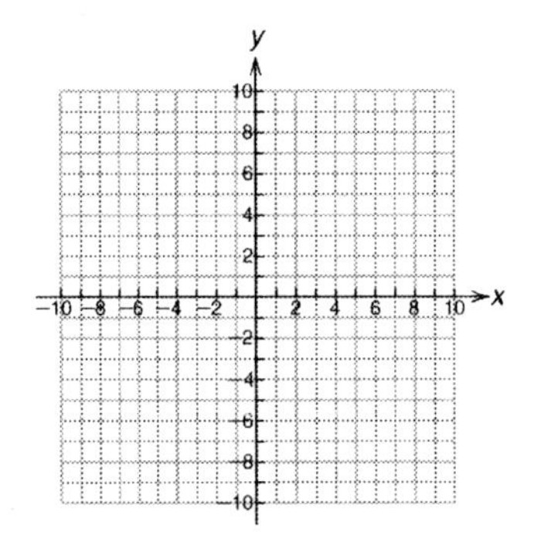

Name: Date:
Instructor: Section:

15. $(8, -6), (0, -2)$

15.______________

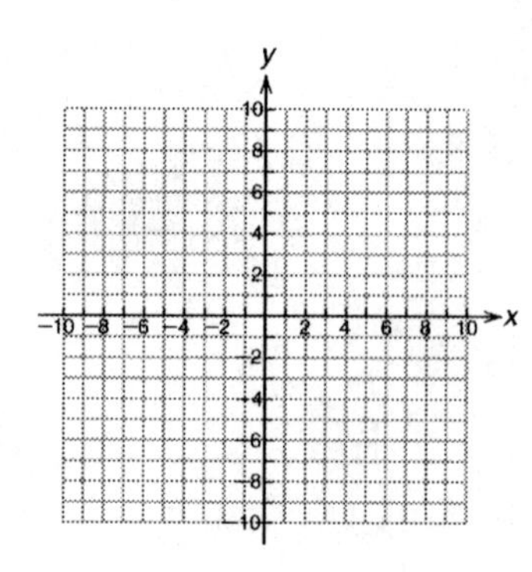

Objective 3 Find a linear equation to describe real data.

Solve.

16. It costs \$14.50 to make 75 copies of a document. It costs \$20.50 to make 175 copies of the same document.

a. Find a linear equation for the cost y to make x copies of the document.

b. How much will it cost to make 500 copies of the document?

16a.______________

b.______________

17. When Millie called Pete's Plumbing, Pete worked 5 hours and charged Millie \$250. When Rosalee called Pete, he worked 7 hours and charged \$310.

a. Find a linear equation for Pete's rate y for working x hours.

b. How much would Pete charge Fred for working 10 hours?

17a.______________

b.______________

Name: Date:
Instructor: Section:

Objective 4 Find the equation of a parallel or perpendicular line.

Find the slope-intercept form of the equation of a line that meets the given conditions.

18. Parallel to $y=-2$, through $(4,\ -3)$ **18.**________________

19. Parallel to $x=1$, through $(-9,\ 13)$ **19.**________________

20. Parallel to $x=-5$, through $(-6,\ -4)$ **20.**________________

21. Parallel to $y=5$, through $(2,\ 3)$ **21.**________________

22. Perpendicular to $y=3x-7$, through $(-8,2)$ **22.**________________

23. Perpendicular to $x=5$, through $(4,-4)$ **23.**________________

Name: Date:
Instructor: Section:

Find the slope-intercept form of the equation of a line that is parallel to the graphed line and that passes through the point plotted on the graph.

24.

24. 

Find the slope-intercept form of the equation of a line that is perpendicular to the graphed line and that passes through the point plotted on the graph.

25.

25. 

26.

26. 

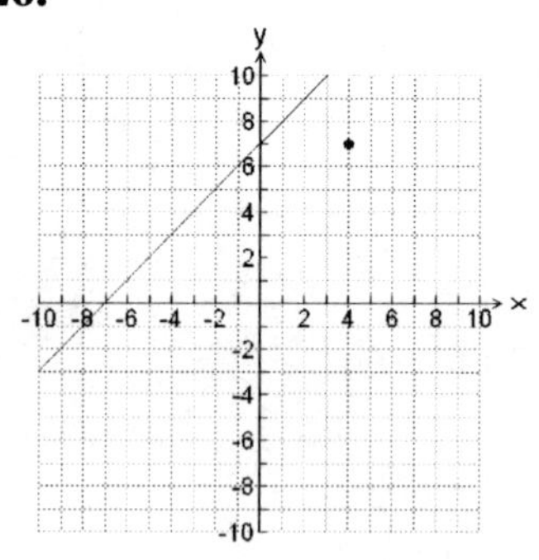

Name: Date:
Instructor: Section:

Chapter 2 GRAPHING LINEAR EQUATIONS

2.4 Linear Inequalities

Learning Objectives
1 Determine whether an ordered pair is a solution of a linear inequality in two variables.
2 Graph a linear inequality in two variables.
3 Graph a linear inequality involving a horizontal or vertical line.
4 Graph linear inequalities associated with applied problems.

Key Terms
Use the most appropriate term or phrase from the given list to complete each statement in exercises 1-2.

weak **strict** **test point** **boundary point** **solid** **dashed**

1. When graphing a linear inequality, use a ________________ to determine which side of the line to shade.

2. Inequalities that involve the symbols < or > are called ________________ inequalities.

Objective 1 Determine whether an ordered pair is a solution of a linear inequality in two variables.

Determine whether the ordered pair is a solution to the given linear inequality.

1. $-5x+4y>6$, $(8,19)$ **1.**________________

2. $y<-3x-6$, $(-2,2)$ **2.**________________

3. $y<-4x-4$, $(-9,3)$ **3.**________________

Name: Date:
Instructor: Section:

4. $-9x+8y>5$, $(9,17)$ **4.**_______________

Objective 2 Graph a linear inequality in two variables.

Complete the solution of the linear inequality by shading the appropriate region.

5. $y \geq -3$ **5.**

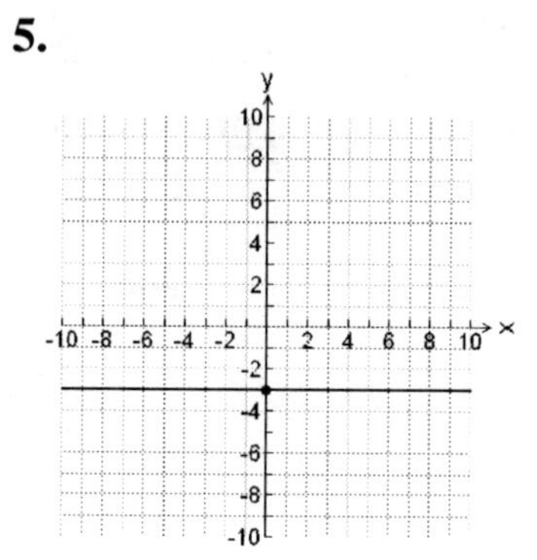

6. $x < -4$ **6.**

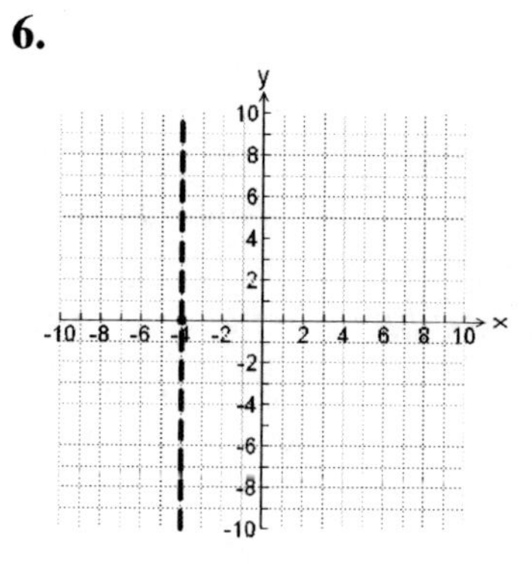

Which graph, A or B, represents the solution to the linear inequality?

7. $y < -4$ **7.**_______________

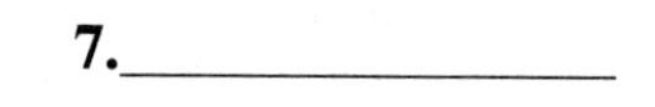

A)

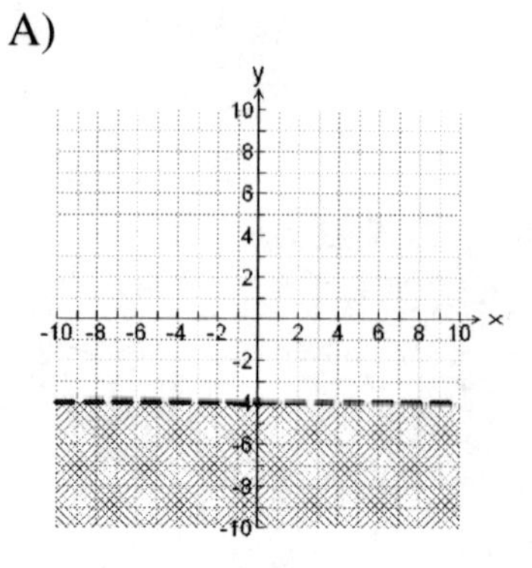

B)

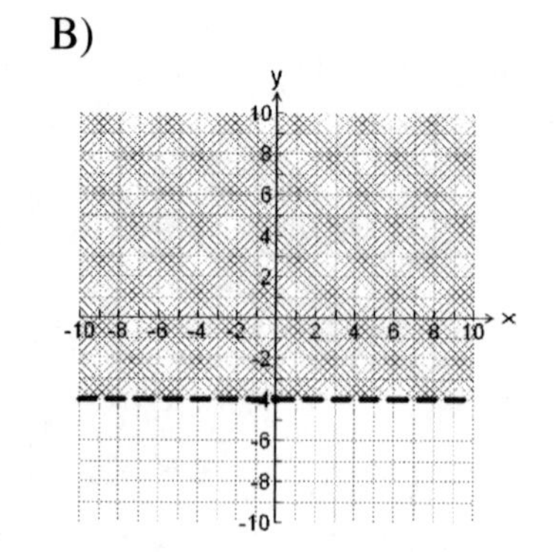

Name: Date:
Instructor: Section:

8. $x > -2$ **8.**________________

A)

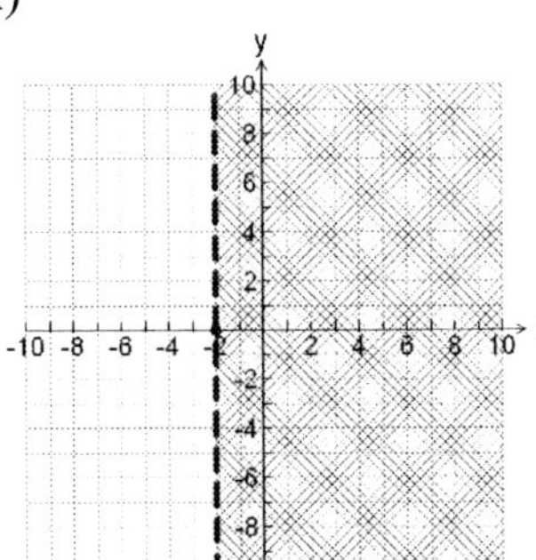

B)

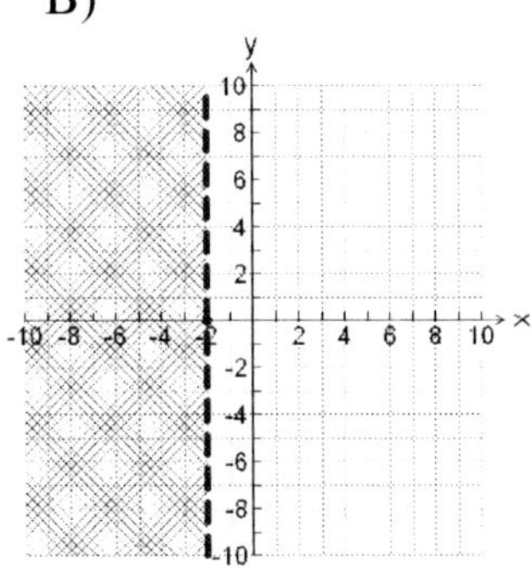

Determine the missing inequality sign (<, >, $\leq$, or $\geq$) for the linear inequality based on the given graph.

9. y ______ $-\frac{3}{2}x+9$ **9.**________________

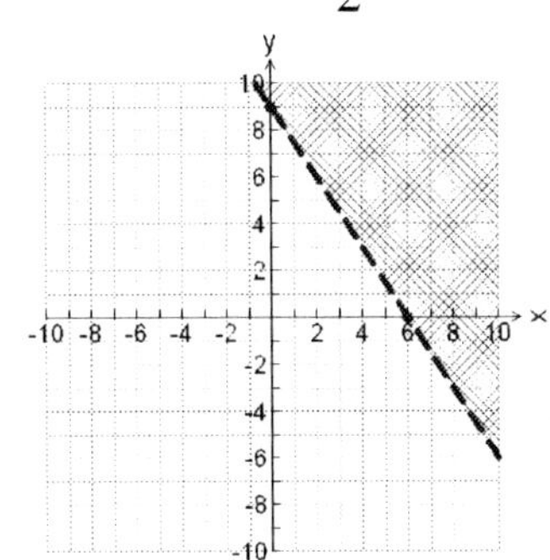

10. y ______ -7 **10.**________________

Name: Date:
Instructor: Section:

Graph the linear inequality.

11. $y \geq 4x$

11.

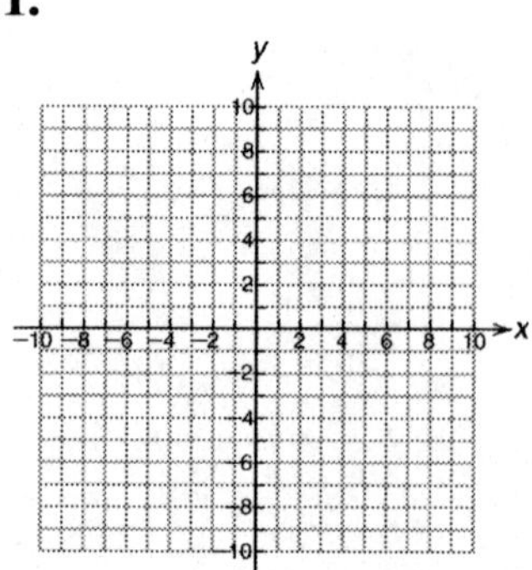

12. $y < -2x + 5$

12.

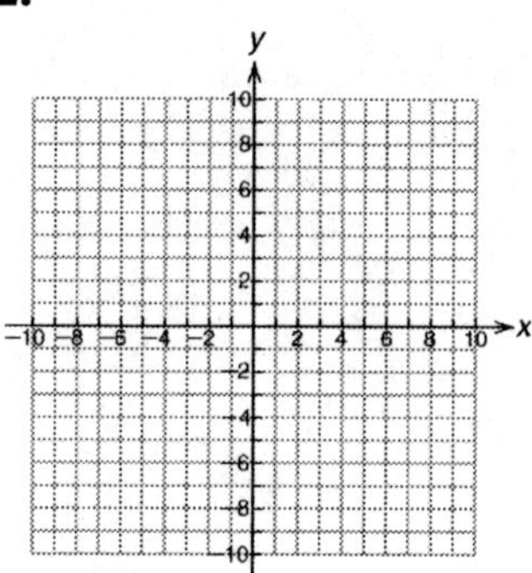

13. $y > 3x - 1$

13.

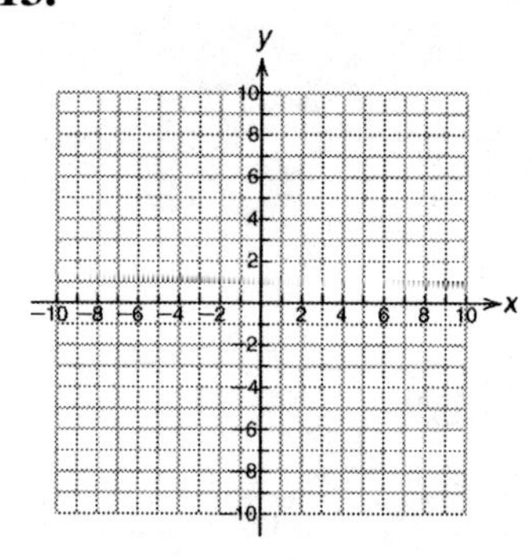

14. $y \leq -2x - 7$

14.

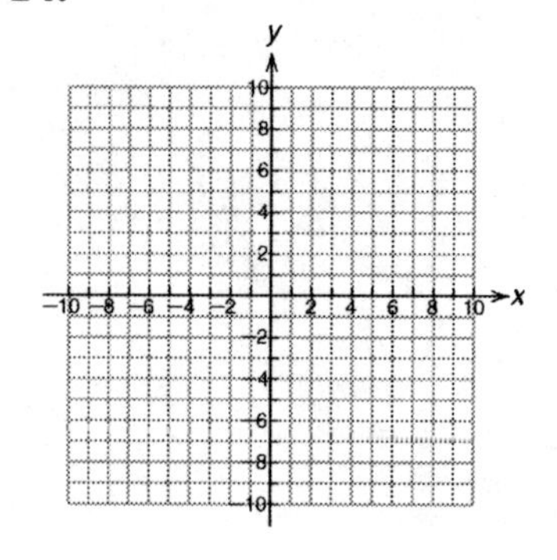

Name: Date:
Instructor: Section:

Objective 3 Graph a linear inequality involving a horizontal or vertical line.

Graph the linear inequality.

15. $y < -3$

15.

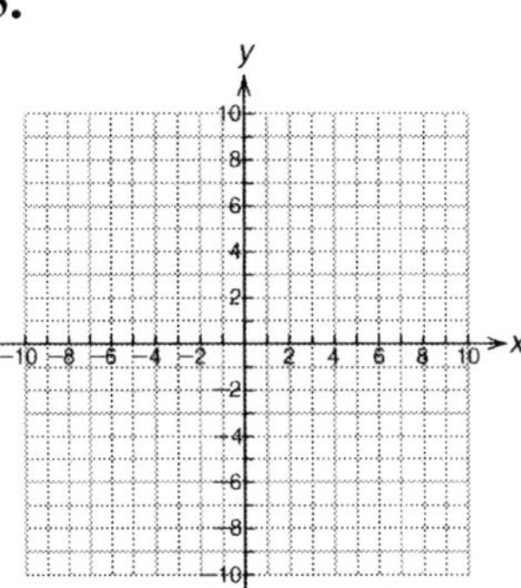

16. $y > 6$

16.

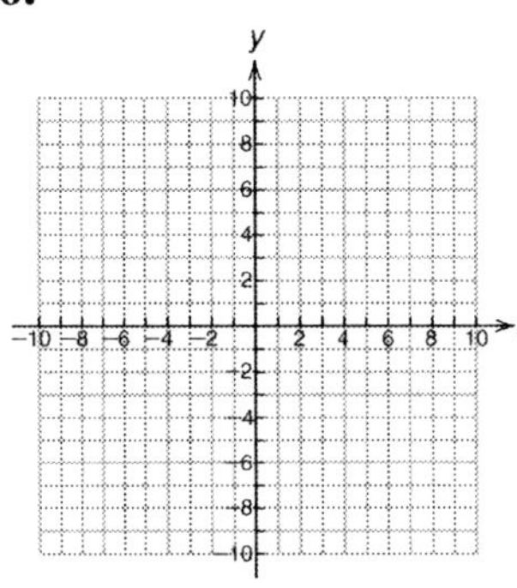

17. $x < 8$

17.

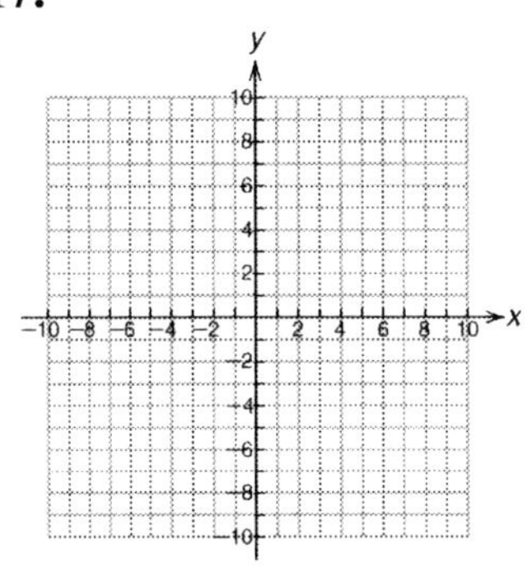

18. $x \geq -2$

18.

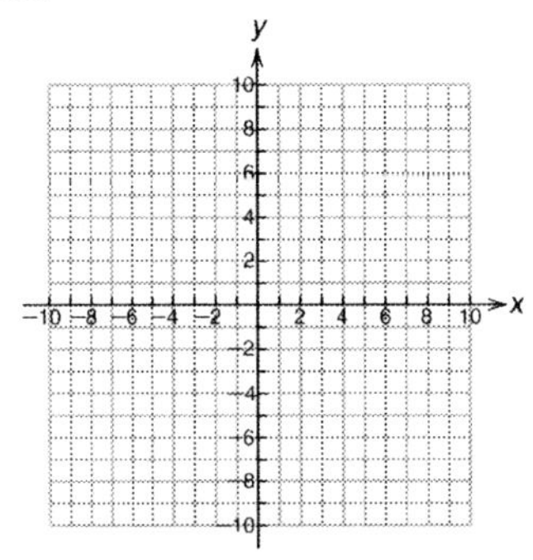

Name: Date:
Instructor: Section:

Objective 4 Graph linear inequalities associated with applied problems.

Solve.

19. The CEO of TM Solutions has determined that to remain in business TM must ship at least 1400 units monthly to its two distributors.
a. Set up and graph an inequality involving the number of units TM must ship to its distributors.
b. Find two ordered pairs that are in the solution set. (Answers will vary.)

19a.________________

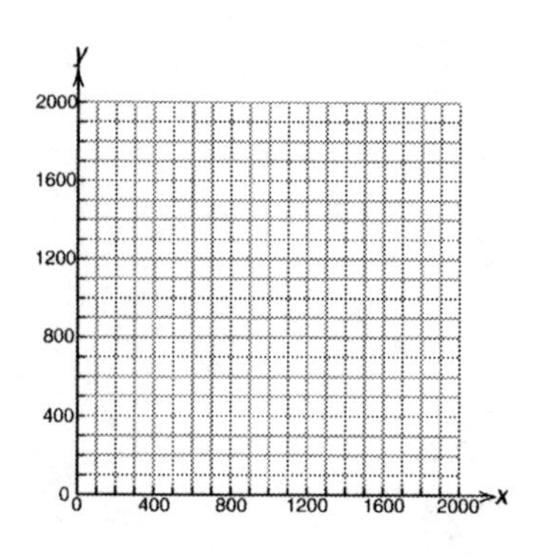

b.________________

Determine the linear inequality associated with the given solution. (First, find the equation of the line. Then rewrite this equation as an inequality with the appropriate inequality sign: <, >, $\leq$, or $\geq$.)

20.

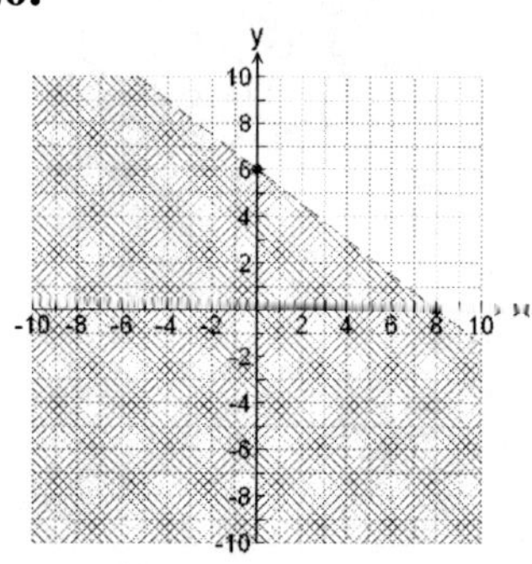

20.________________

21.

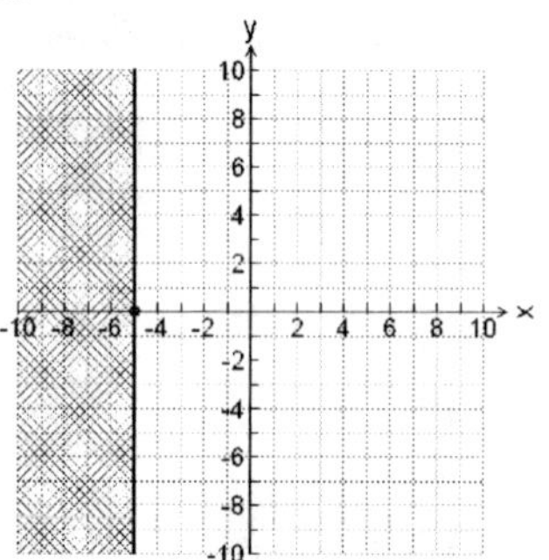

21.________________

Name: Date:
Instructor: Section:

Find the region that contains ordered pairs that are solutions to both inequalities. First graph each inequality separately, and then shade the region that the two graphs have in common.

22. $y \geq 4x - 7$ and $y \leq 8$

22.

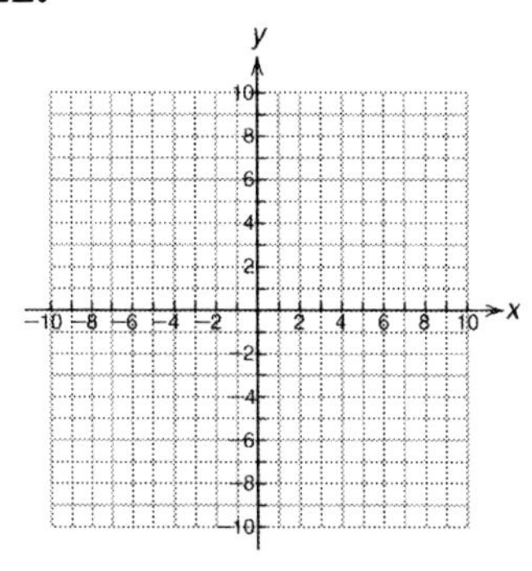

Name: Date:
Instructor: Section:

Chapter 2 GRAPHING LINEAR EQUATIONS

2.5 Linear Functions

Learning Objectives

1 Define and understand function, domain, and range.
2 Evaluate functions.
3 Graph linear functions.
4 Interpret the graph of a linear function.
5 Determine the domain and range of a function from its graph.
6 Find a linear function that meets given conditions.

Key Terms

Use the most appropriate term from the given list to complete each statement in exercises 1-3.

relation **function** **domain** **range** **linear** **constant**

1. A ________________ function has the same value, regardless of the input value.

2. A ________________ is a rule that takes an input value from one set and assigns to it an output value from another set.

3. The ________________ of a function is the set of output values.

Objective 1 Define and understand function, domain, and range.

Solve.

1. Determine whether a function exists with set A as the input and set B as the output and whether a function exists with set B as the input and set A as the output.

1.________________

Set A Person	Set B Years of Service
Jay Delsinger	15
Meredith Walker	12
Jeff Wills	9
Anita Morales	11
Sung Kim	7

Name: Date:
Instructor: Section:

For the given set of ordered pairs, determine whether a function could be defined for which the input would be an x-coordinate and the output would be the corresponding y-coordinate. If a function cannot be defined in this matter, explain why..

2. $\{(9,1),(-7,-5),(2,-1),(4,-8)\}$ **2.**________________

3. $\{(7,3),(-8,-5),(2,-2),(-8,-9)\}$ **3.**________________

4. $\{(7,4),(-9,4),(1,-1),(5,-7)\}$ **4.**________________

Objective 2 Evaluate functions.

Evaluate the given function.

5. $g(x)=x+8,\ g(3)$ **5.**________________

6. $P(x)=4x,\ P(9)$ **6.**________________

7. $h(x)=4x+15,\ h\left(\frac{7}{2}\right)$ **7.**________________

8. $h(x)=-6x-5,\ h\left(-\frac{5}{3}\right)$ **8.**________________

Name: Date:
Instructor: Section:

Objective 3 Graph linear functions.

Graph the linear function.

9. $f(x) = 4x + 5$

9.

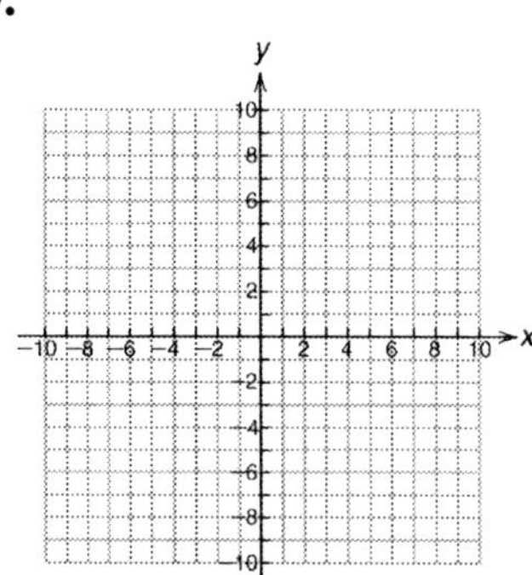

10. $f(x) = -3x - 4$

10.

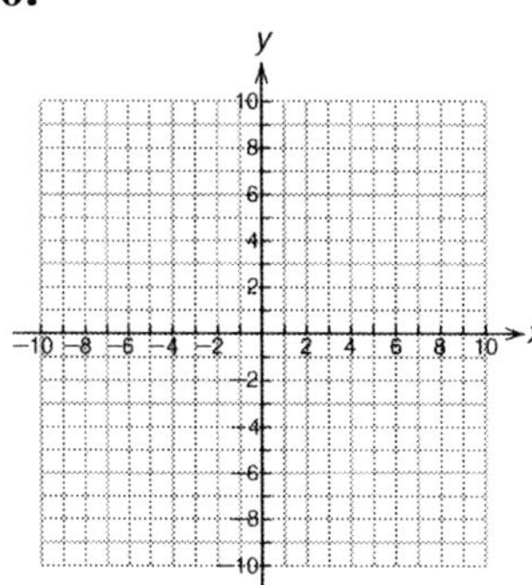

11. $f(x) = 4x - 16$

11.

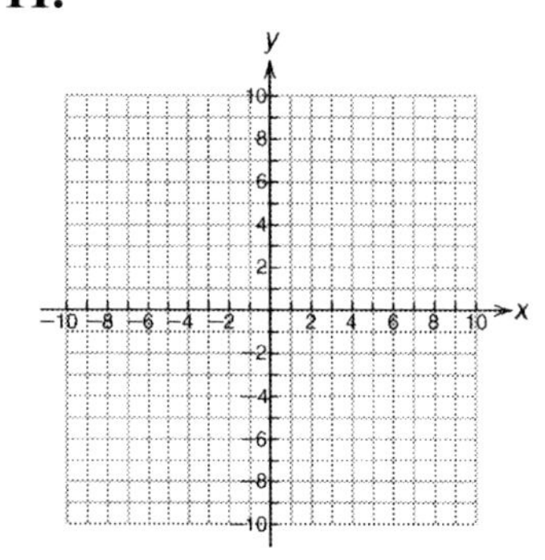

12. $f(x) = -8x - 24$

12.

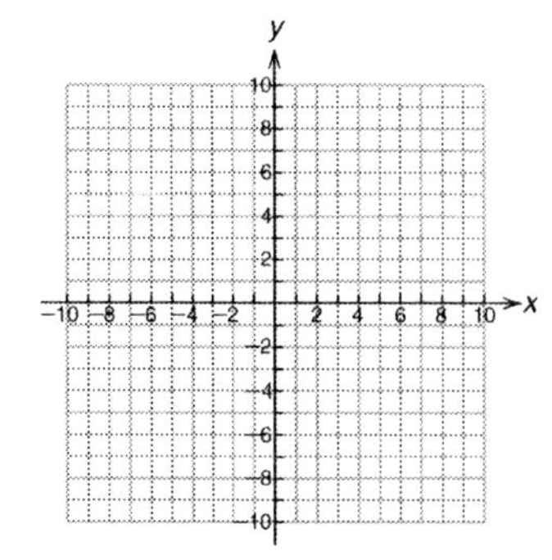

Name: Date:
Instructor: Section:

13. $f(x)=x-7$

13.

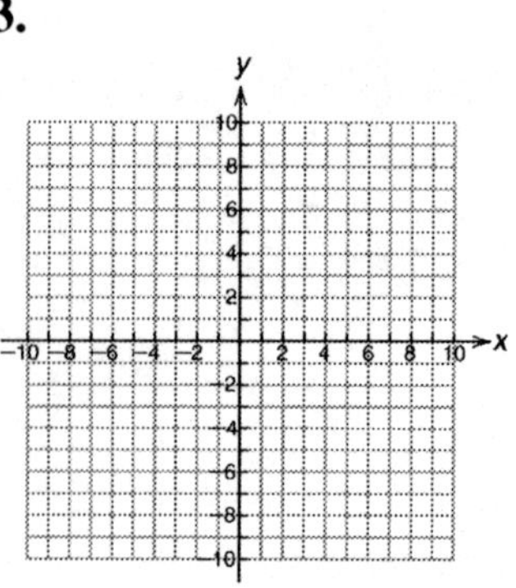

Objective 4 Interpret the graph of a linear function.

Solve.

14. Refer to the graph of the function $f(x)$.

a. Find the x-intercept.
b. Find the y-intercept.
c. Find $f(5)$.
d. Find a value a such that $f(a)=8$.

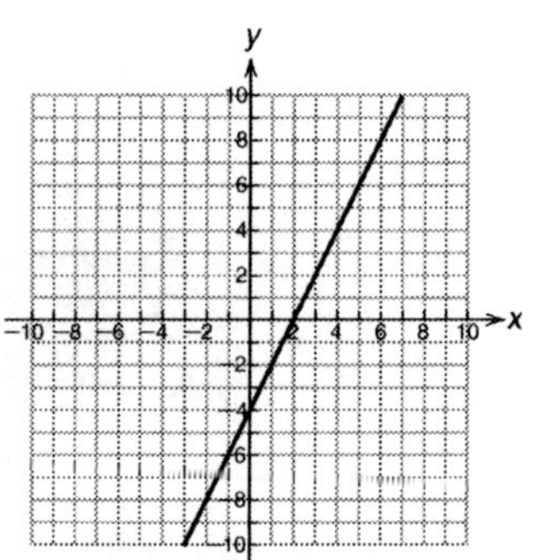

14a.________________

b.________________

c.________________

d.________________

15. Refer to the graph of the function $f(x)$.

a. Find the x-intercept.
b. Find the y-intercept.
c. Find $f(-3)$.
d. Find a value a such that $f(a)=4$.

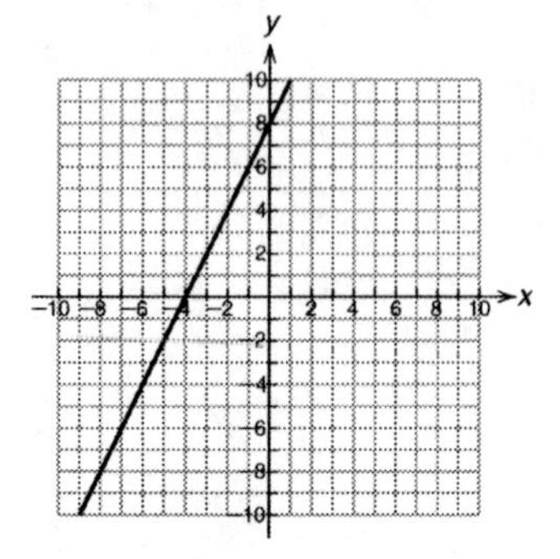

15a.________________

b.________________

c.________________

d.________________

Name: Date:
Instructor: Section:

Objective 5 Determine the domain and range of a function from its graph.

Find the domain and range of the given function.

16. 16.________________

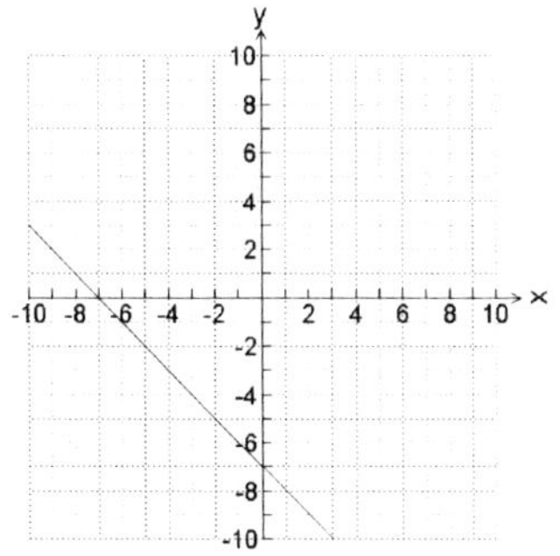

17. 17.________________

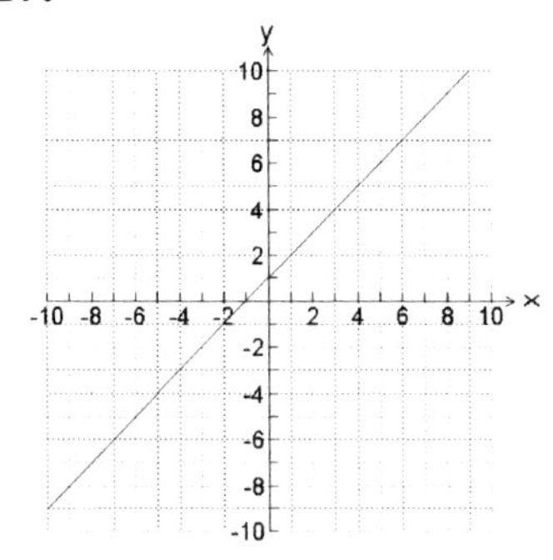

18. 18.________________

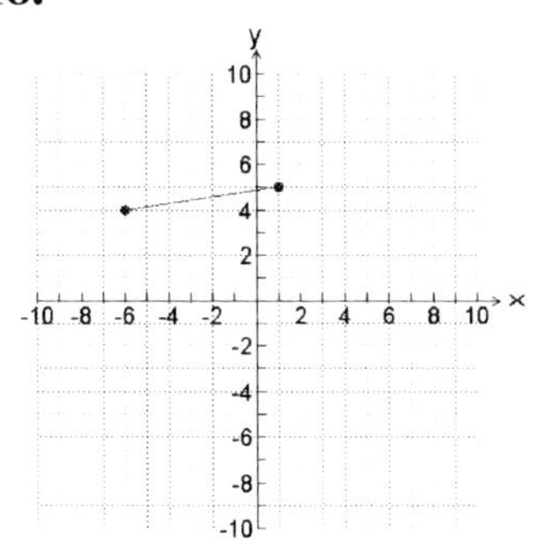

19. 19.________________

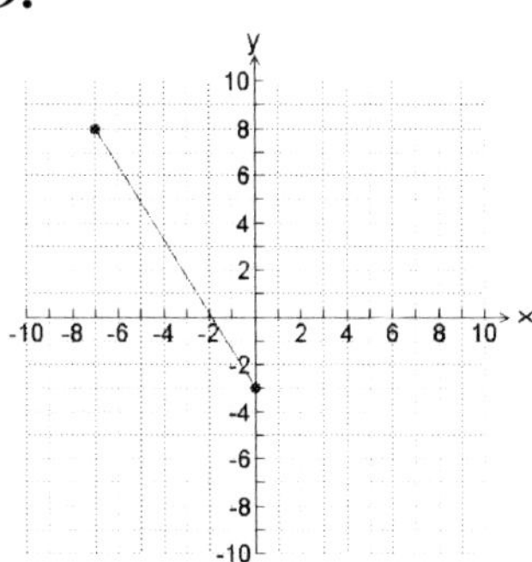

Name: Date:
Instructor: Section:

20.

20.

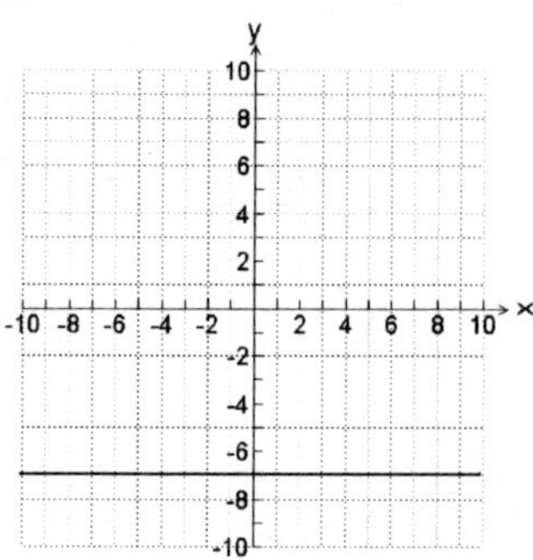

Objective 6 Find a linear function that meets given conditions.

Create a linear function f(x) whose graph has the given slope and y-intercept.

21. Slope of 3, *y*-intercept at $\left(0,\ \frac{1}{2}\right)$ **21.**________________

22. Slope of -5, *y*-intercept at $(0,\ -7)$ **22.**________________

Create a linear function f(x), whose graph has the given slope, that meets the given condition.

23. Slope of $\frac{1}{5}$, $f(10)=-2$ **23.**________________

24. Slope of 1.2, $f(-2)=-5.7$ **24.**________________

Create a linear function f(x) that meets the given condition.

25. $f(3)=2$, $f(-2)=7$ **25.**________________

26. $f(-1)=-10$, $f(4)=5$ **26.**________________

Name: Date:
Instructor: Section:

Chapter 2 GRAPHING LINEAR EQUATIONS

2.6 Absolute Value Functions

Learning Objectives
1 Graph absolute value functions.
2 Determine an absolute value function from its graph.

Objective 1 Graph absolute value functions.

Graph each absolute value function. State the domain and range of each function.

1. $f(x)=|x+6|+6$

1.________________

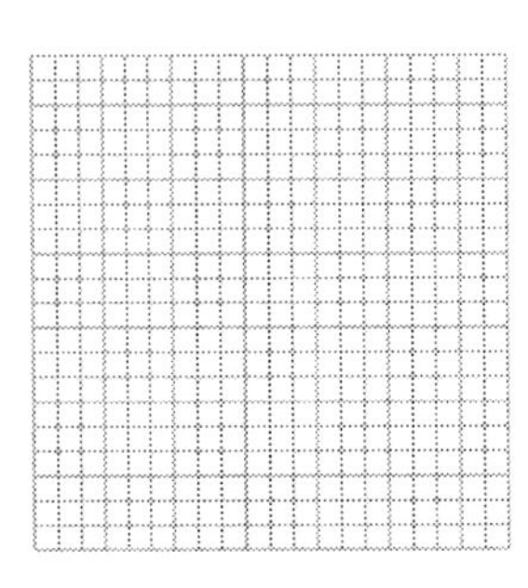

2. $f(x)=-|2x+7|-4$

2.________________

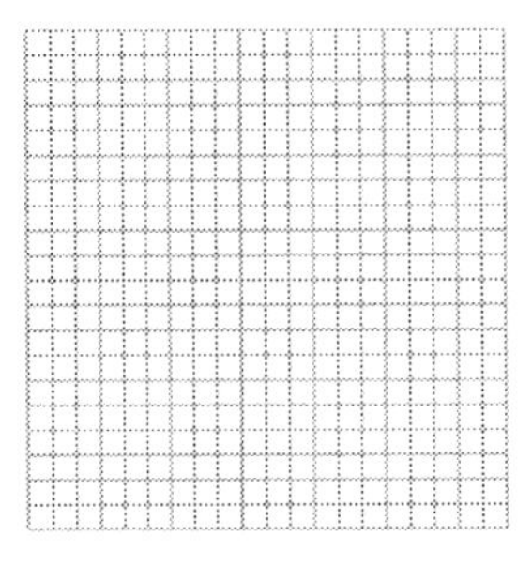

3. $f(x)=|3x+4|-2$

3.________________

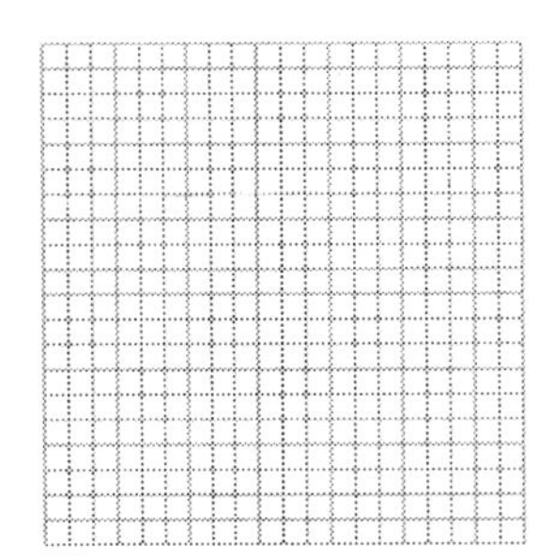

Name: Date:
Instructor: Section:

4. $f(x) = |x| - 4$

4.________________

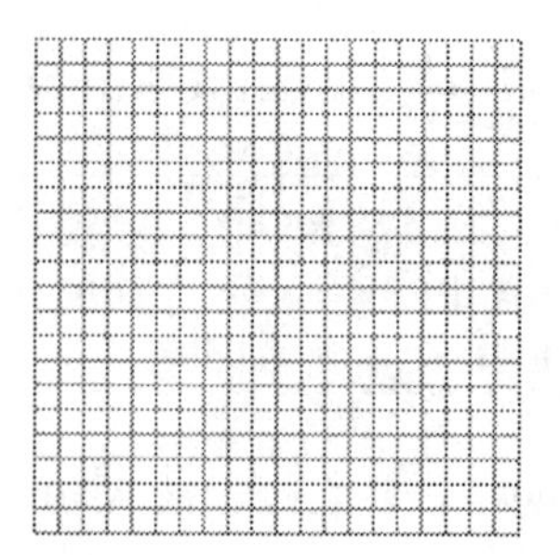

Find the intercepts of each absolute value function.

5. $f(x) = |x-3| - 6$

5.________________

6. $f(x) = |x+6|$

6.________________

7. $f(x) = |x-4|$

7.________________

8. $f(x) = |x+2| + 5$

8.________________

Name: Date:
Instructor: Section:

Use the graph of the absolute value function f(x) to solve the inequality.

9. $f(x) \leq 3$ **9.**________________

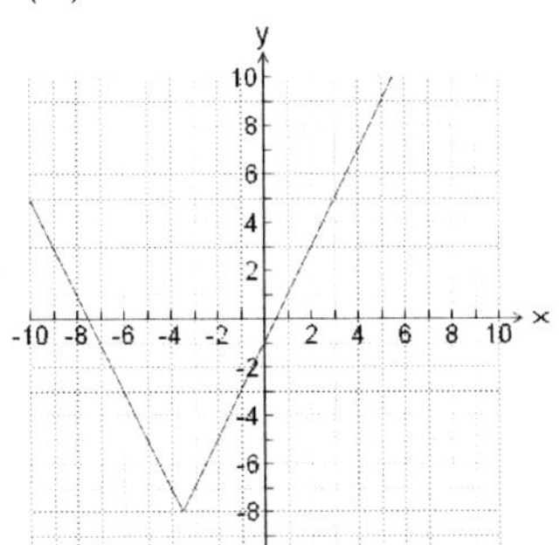

Objective 2 Determine an absolute value function from its graph.

Determine the absolute value function f(x) that has been graphed.

10. **10.**________________

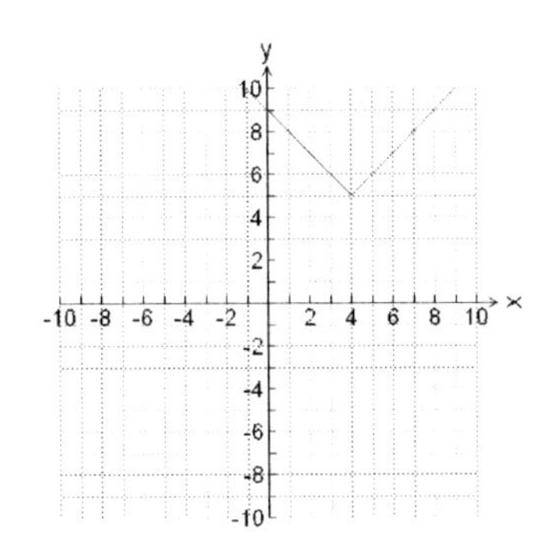

11. **11.**________________

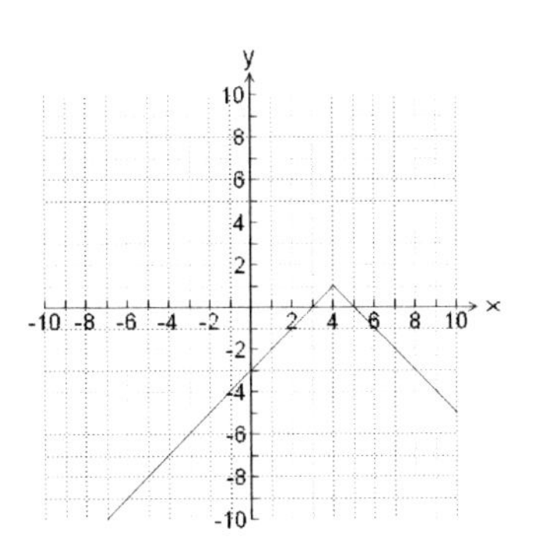

Name: Date:
Instructor: Section:

12. **12.**_______________

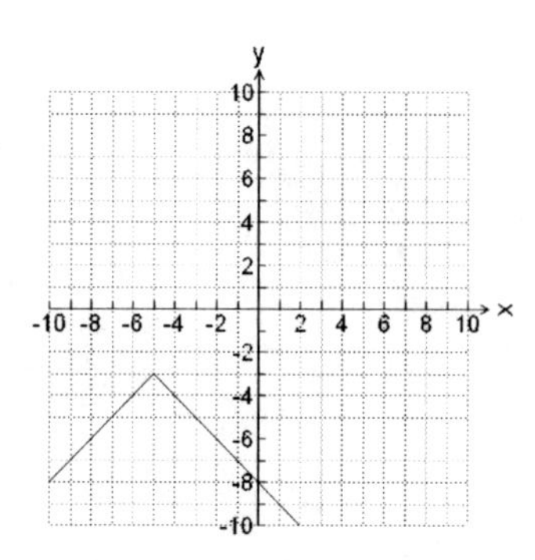

13. **13.**_______________

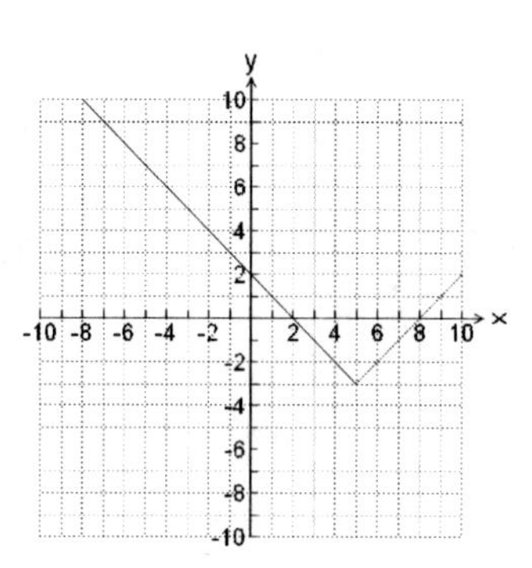

14. **14.**_______________

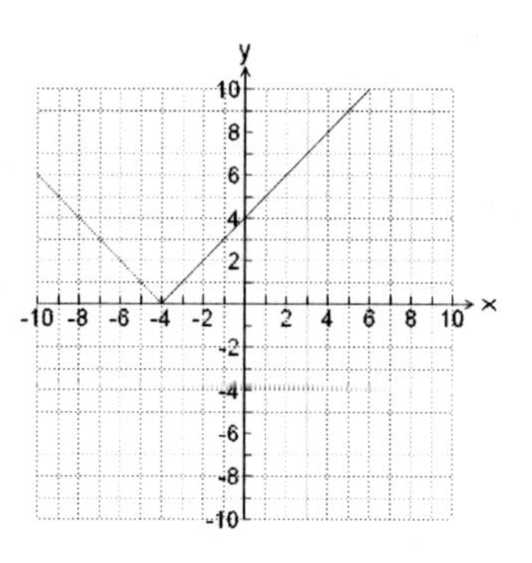

15. **15.**_______________

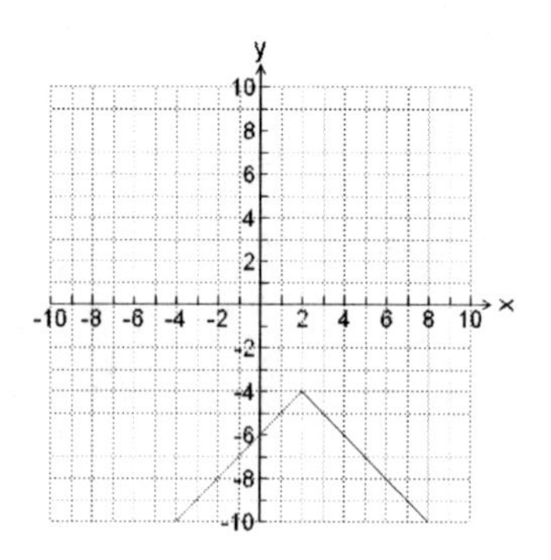

Name: Date:
Instructor: Section:

Chapter 3 SYSTEMS OF EQUATIONS

3.1 Systems of Two Linear Equations in Two Unknowns

Learning Objectives

1. Determine whether an ordered pair is a solution to a system of two equations in two unknowns.
2. Solve a system of two linear equations in two unknowns graphically.
3. Solve a system of two linear equations using the substitution method.
4. Solve a system of two linear equations using the addition method.

Key Terms

Use the most appropriate term or phrase from the given list to complete each statement in exercises 1-3.

independent **inconsistent** **dependent**

1. The graph of a(n) ____________________ system of two equations in two unknowns consists of two parallel lines.

2. The graph of a(n) ____________________ system of two equations in two unknowns is a single line.

3. The graph of a(n) ____________________ system of two equations in two unknowns is a pair of lines that intersect in exactly one point.

Objective 1 Determine whether an ordered pair is a solution to a system of two equations in two unknowns.

Is the ordered pair a solution of the given system of equations?

1. $(5,4)$, $\begin{aligned} 4x-y&=24 \\ 9x-3y&=33 \end{aligned}$

1. ____________

2. $(7,2)$, $\begin{aligned} 7x-y&=47 \\ 4x-5y&=18 \end{aligned}$

2. ____________

Name: Date:
Instructor: Section:

3. $(5,7)$, $\begin{aligned} 6x - y &= 37 \\ 8x - 8y &= -16 \end{aligned}$

3.________________

4. $(2,8)$, $\begin{aligned} 8x - y &= 24 \\ 8x - 4y &= -16 \end{aligned}$

4.________________

Objective 2 Solve a system of two linear equations in two unknowns graphically.

Solve the system by graphing. If the system is inconsistent and has no solution, state this. If the system is dependent, write the form of the solution for any real number x.

5. $\begin{aligned} y &= x - 5 \\ y &= -3x + 1 \end{aligned}$

5.________________

6. $\begin{aligned} y &= \frac{1}{2}x + 3 \\ y &= -2x - 7 \end{aligned}$

6.________________

7. $\begin{aligned} 8x - 10y &= -18 \\ -12x + 15y &= 27 \end{aligned}$

7.________________

Name: Date:
Instructor: Section:

8. $16x - 9y = 8$

$y = \frac{8}{3}x + \frac{4}{3}$

8.______________

9. $-4x + 3y = 9$

$7x + 4y = 12$

9.______________

10. $\frac{1}{6}x - \frac{1}{4}y = -\frac{4}{3}$

$y = 6$

10.______________

Objective 3 Solve a system of two linear equations using the substitution method.

Solve the system by substitution. If the system is inconsistent and has no solution, state this. If the system is dependent, write the form of the solution for any real number x.

11. $x = -1$

$5x - 2y = -1$

11.______________

Name: Date:
Instructor: Section:

12. $\begin{aligned} -x+3y&=-21 \\ 4x-y&=51 \end{aligned}$

12.________________

13. $\begin{aligned} x+y&=7 \\ y&=x-5 \end{aligned}$

13.________________

14. $\begin{aligned} 7x+5y&=-41 \\ x&=7-5y \end{aligned}$

14.________________

15. $\begin{aligned} x+8y&=21 \\ 9x+4y&=-15 \end{aligned}$

15.________________

16. $\begin{aligned} x-y&=-2 \\ 8x+2y&=-36 \end{aligned}$

16.________________

Name: Date:
Instructor: Section:

17. $6x-3y=-30$
$3x-y=-17$

17.________________

18. $40x-5y=25$
$8x-y=5$

18.________________

Objective 4 Solve a system of two linear equations using the addition method.

Solve the system by addition. If the system is inconsistent and has no solution, state this. If the system is dependent, write the form of the solution for any real number x.

19. $8x+6y=-50$
$5x-6y=-41$

19.________________

20. $5x+6y=5$
$10x+12y=10$

20.________________

21. $3x-9y=3$
$3x-9y=4$

21.________________

Name: Date:
Instructor: Section:

22. $\begin{aligned} 5x-2y&=4 \\ 2x+5y&=48 \end{aligned}$

22.________________

23. $\begin{aligned} \frac{7}{2}x+2y&=-1 \\ \frac{1}{4}x+\frac{1}{4}y&=2 \end{aligned}$

23.________________

24. $\begin{aligned} \frac{3}{7}x+\frac{4}{7}y&=4 \\ \frac{1}{5}x+\frac{2}{15}y&=\frac{4}{3} \end{aligned}$

24.________________

25. $\begin{aligned} x+y&=51 \\ 0.05x+0.25y&=6.15 \end{aligned}$

25.________________

Name: Date:
Instructor: Section:

Chapter 3 SYSTEMS OF EQUATIONS

3.2 Applications of Systems of Equations

Learning Objectives

1. Solve applied problems involving systems of equations by using the substitution method or the addition method.
2. Solve geometry problems using a system of equations.
3. Solve interest problems using a system of equations.
4. Solve mixture problems using a system of equations.
5. Solve motion problems using a system of equations.

Objective 1 Solve applied problems involving systems of equations by using the substitution method or the addition method.

Solve.

1. Woodstock's Pizza sold 37 more large pizzas last Friday night than they sold medium pizzas. The number of large pizzas sold was 21 less than 3 times the number of medium pizzas sold. How many large pizzas did they sell, and how many medium pizzas did they sell?

1._______________

2. A piece of pipe that was 96 inches long was cut into two pieces. One piece is 17 inches longer than the other piece. How long is each piece?

2._______________

Name: Date:
Instructor: Section:

3. Last week a school play was attended by 28 adults and 55 children who paid a total of $153. This week the play was attended by 60 adults and 95 children who paid a total of $305. How much did the school charge each adult and how much did they charge each child?

3.________________

4. Ramon bought 10 cups of coffee and 8 muffins for $28.30. Laverne bought 5 cups of coffee and 12 muffins for $24.95. What is the price of each cup of coffee and each muffin?

4.________________

5. Tina's purse contains $5 bills and $10 bills. If the purse contains 39 bills worth a total of $355, how many $5 bills are there?

5.________________

6. The photography club collected $480 by selling t-shirts and baseball hats. They charged $8 for each t-shirt and $10 for each baseball hat. If they sold a total of 52 items, how many t-shirts did they sell?

6.________________

Name: Date:
Instructor: Section:

7. A basketball team played 72 games. They won 28 more than they lost. How many games did they win?

7.________________

8. Bob took a test in which multiple-choice questions are worth 10 points and essay questions are worth 15 points. He answered 14 questions correctly, and his score was 170 points. Find the number of questions of each type he answered correctly.

8.________________

9. A parking meter contains nickels and quarters worth $11.90. There are 78 coins in all. Find how many of each there are.

9.________________

10. At a concession stand, five hot dogs and two hamburgers cost $10.75; two hot dogs and five hamburgers cost $13.75. Find the cost of one hot dog and the cost of one hamburger.

10.________________

Name: Date:
Instructor: Section:

Objective 2 Solve geometry problems by using a system of equations.

Solve.

11. Two angles are complementary. The measure of one angle is 42° less than the measure of the other angle. Find the measures of each angle.

11.________________

12. Two angles are complementary. The measure of one angle is 4 times the measure of the other angle. Find the measures of each angle.

12.________________

13. Two angles are supplementary. The measure of one angle is 49° less than the measure of the other angle. Find the measures of each angle.

13.________________

14. Two angles are supplementary. The measure of one angle is double the measure of the other angle. Find the measures of each angle.

14.________________

Name: Date:
Instructor: Section:

15. Two angles are supplementary. One is 75° more than twice the other. Find the measures of the angles.

15.________________

16. The perimeter of a standard-sized rectangular rug is 44 feet. The width is 2 feet less than the length. Find the dimensions.

16.________________

17. The perimeter of a rectangle is 122 meters. The length is 4 meters more than twice the width. Find the dimensions.

17.________________

Objective 3 Solve interest problems by using a system of equations.

Solve.

18. Tamyra invested a total of $3500 in two mutual funds. One fund made a 9% profit and the other made a 3% profit. If Tamyra earned $15 more from the fund that made a 9% profit than from the fund that made a 3% profit, how much did she invest in each fund?

18.________________

Name: Date:
Instructor: Section:

19. Miriam has CDs at two different banks, making a total deposit of $14,500. One bank pays 5.5% annual interest and the other bank pays 7% interest. In one year Miriam had earned $390 more from the CD that paid 7% interest than she had earned from the other CD. How much was invested in each CD?

19.______________

20. Ms. Brown invested $22,000 in two accounts, one yielding 6% interest and the other yielding 9%. If she received a total of $1680 in interest at the end of the year, how much did she invest in each account?

20.______________

Objective 4 Solve mixture problems by using a system of equations.

Solve.

21. A chemistry student has two solutions. The first solution is 8% alcohol, and the second is 16% alcohol. How many milliliters of each does the student need to mix to produce 56 milliliters of a solution that is 11% alcohol?

21.______________

Name: Date:
Instructor: Section:

22. One canned juice drink is 25% orange juice, another is 5% orange juice. How many liters of each should be mixed together in order to get 20 liters that is 21% orange juice?

22._______________

23. Soybean meal is 18% protein; cornmeal is 9% protein. How many pounds of each should be mixed together in order to get a 360-pound mixture that is 10% protein?

23._______________

24. Pure acid is to be added to a 5% acid solution to obtain 95 liters of 46% solution. What amounts of each should be used?

24._______________

Objective 5 Solve motion problems by using a system of equations.

Solve.

25. An airplane flying against the wind can travel 360 miles in three hours. Flying with the wind, the plane can travel the same distance in two hours. What is the speed of the plane? What is the speed of the wind?

25._______________

Name: Date:
Instructor: Section:

26. Alvin paddled for 4 hours with a 6-km/h current to reach a campsite. The return trip against the same current took 10 hours. Find the speed of the boat in still water.

26._______________

27. An airplane took 4 hours to fly 2400 miles against a headwind. The return trip with the same wind speed took 3 hours. Find the speed of the wind.

27._______________

Name: Date:
Instructor: Section:

Chapter 3 SYSTEMS OF EQUATIONS

3.3 Systems of Linear Inequalities

Learning Objectives

1. Solve a system of linear inequalities by graphing.
2. Solve a system of linear inequalities with more than two inequalities.
3. Graph systems of linear inequalities associated with applied problems.

Objective 1 Solve a system of linear inequalities by graphing.

Graph the system of inequalities.

1. $y > 8x - 3$
$y < -4x + 5$

1.

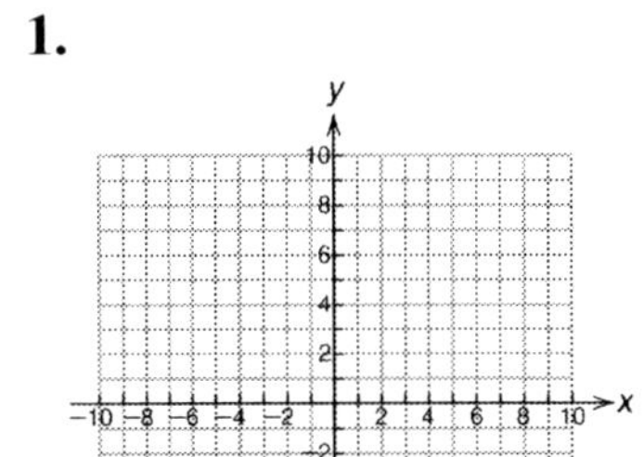

2. $y > 4x + 4$
$y \geq -3x - 3$

2.

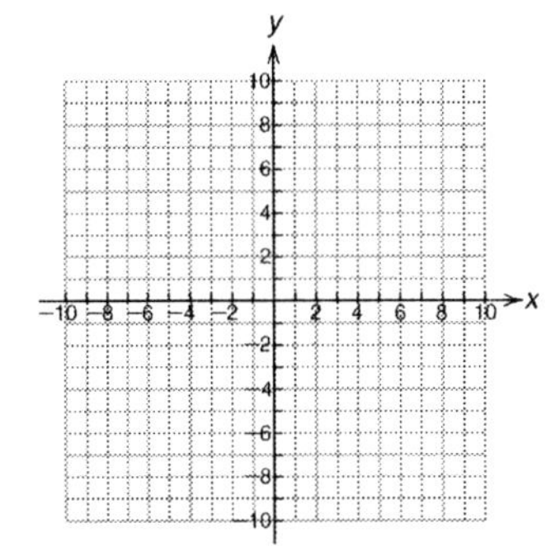

3. $-2x + y < 8$
$2x - y < 2$

3.

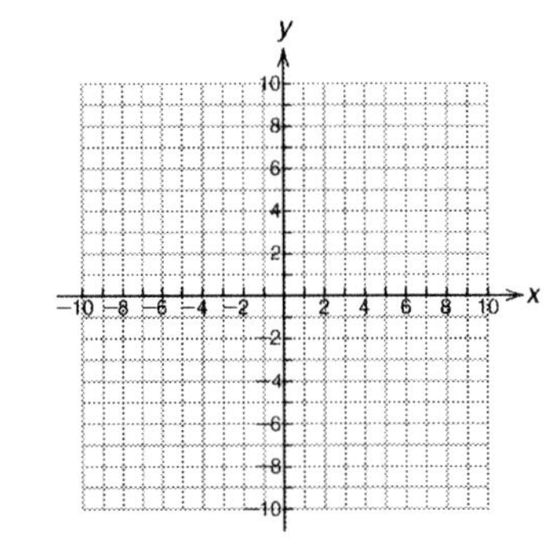

Name: Date:
Instructor: Section:

4. $y < 2x + 2$
$y > 2x - 2$

4.

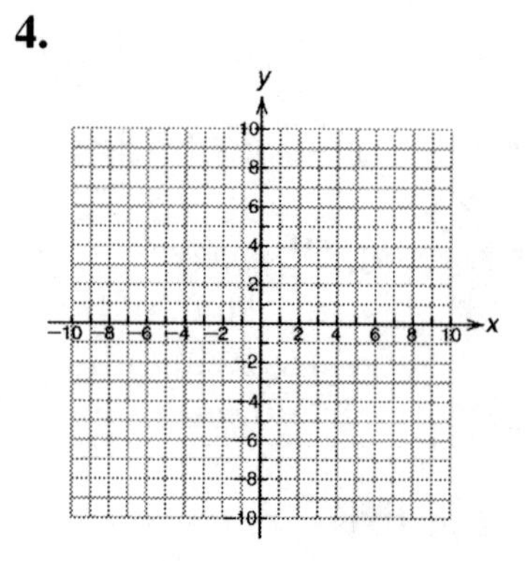

5. $5x + 2y < 10$
$2x - 3y < 6$

5.

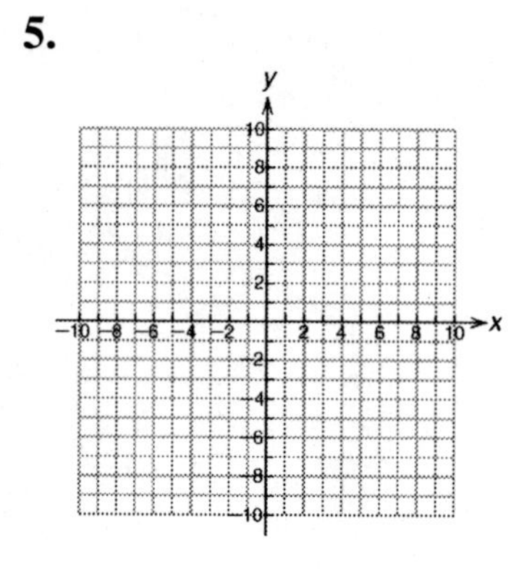

6. $5x + 3y \geq 15$
$3x - 4y \leq 12$

6.

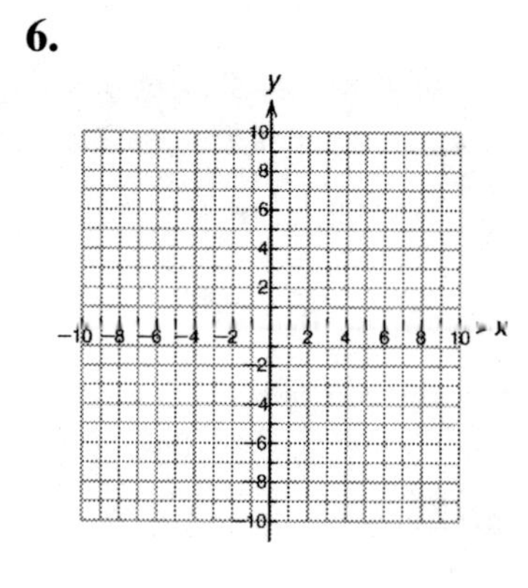

7. $y \geq -2$
$x \geq 3$

7.

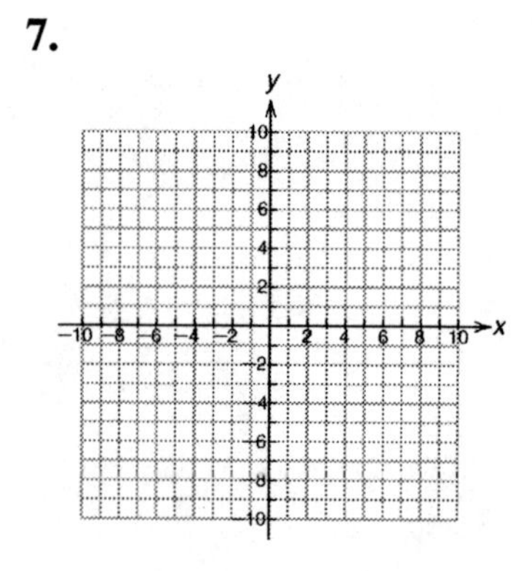

Name: Date:
Instructor: Section:

8. $3x + 2y \geq 6$
$5x - 4y \leq 20$

8.

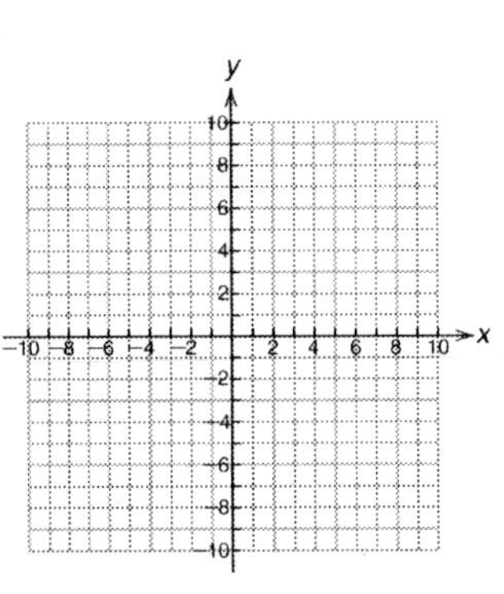

Objective 2 Solve a system of linear inequalities with more than two inequalities.

Graph the system of inequalities.

9. $3x - y \geq -4$
$2x + y < 9$
$y \geq 1$

9.

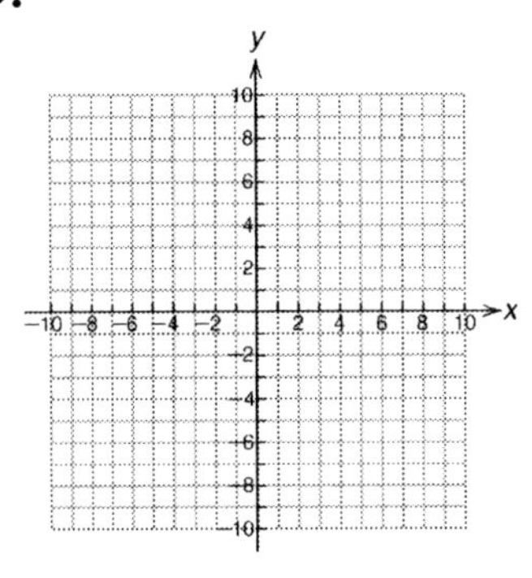

10. $-4x + 3y \geq -7$
$y < 7$
$x > -2$

10.

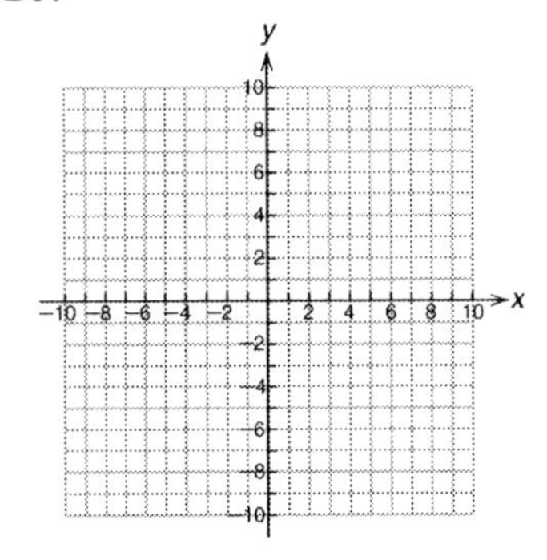

Name: Date:
Instructor: Section:

11. $y < \frac{3}{4}x + 6$

$y < -3x + 8$

$y > -\frac{1}{4}x - 2$

11.

12.

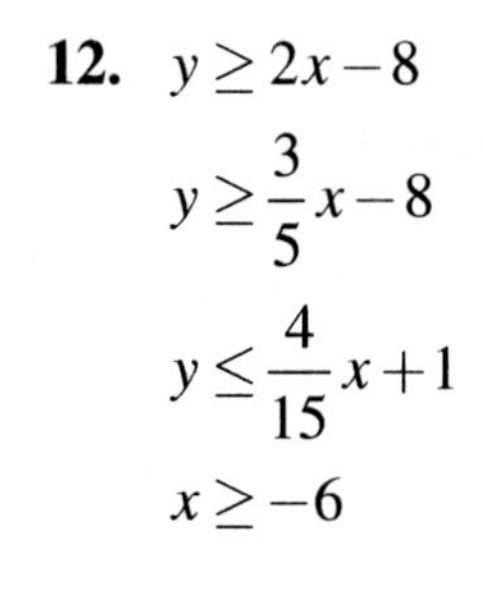

$y \geq 2x - 8$

$y \geq \frac{3}{5}x - 8$

$y \leq \frac{4}{15}x + 1$

$x \geq -6$

12.

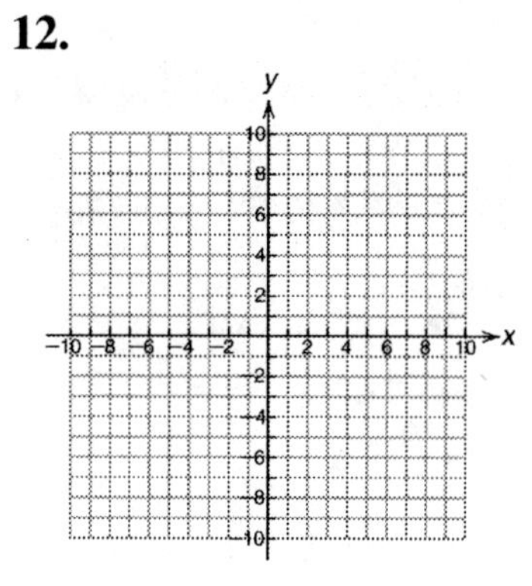

13. $y \leq 2x + 3$

$y \geq -2x + 3$

$x \leq 2$

13.

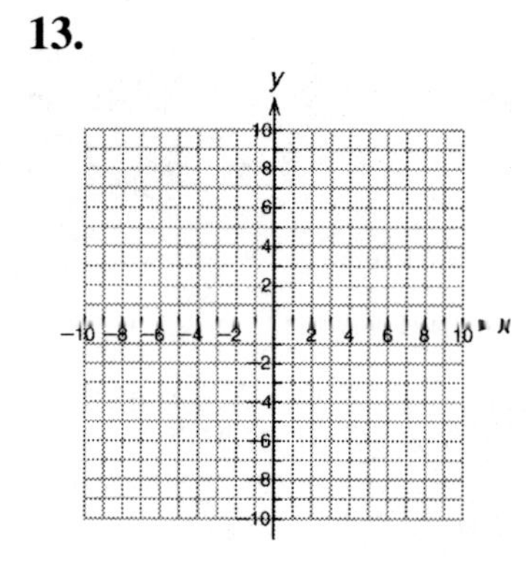

14. $y < 2x + 4$

$y \leq -x$

$y \geq -3$

14.

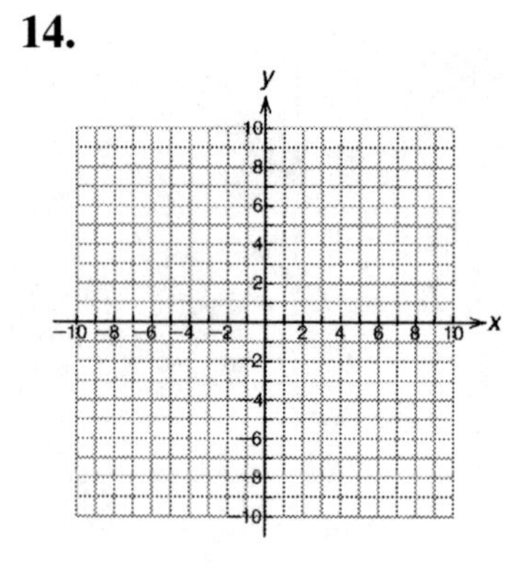

Name: Date:
Instructor: Section:

Objective 3 Graph systems of linear inequalities associated with applied problems.

Solve.

15. A graduate may invite up to 50 people to his graduation party. The number of people invited from school was at least 30 more than the number of people invited from work. Set up and graph a system of inequalities for this situation, letting x represent the number of people who were invited from work and y represent the number of people who were invited from school.

15.

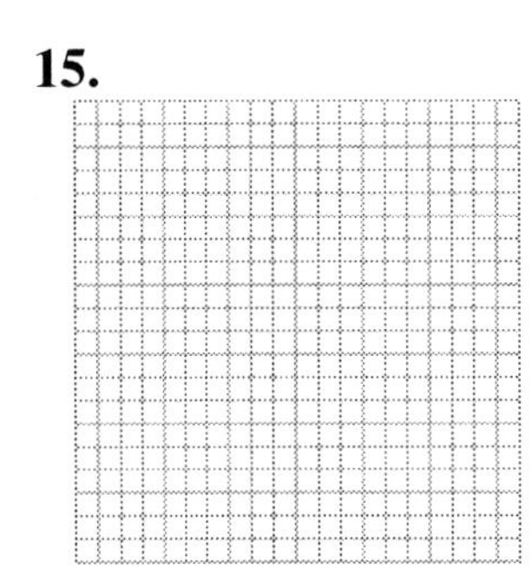

16. An investor has at most $100,000 to divide between two accounts. The amount invested in the second account must be at least 4 times as much as the amount invested in the first account. Set up and graph a system of inequalities for this situation, letting x represent the amount invested in the first account in thousands of dollars and y represent the amount invested in the second account in thousands of dollars.

16.

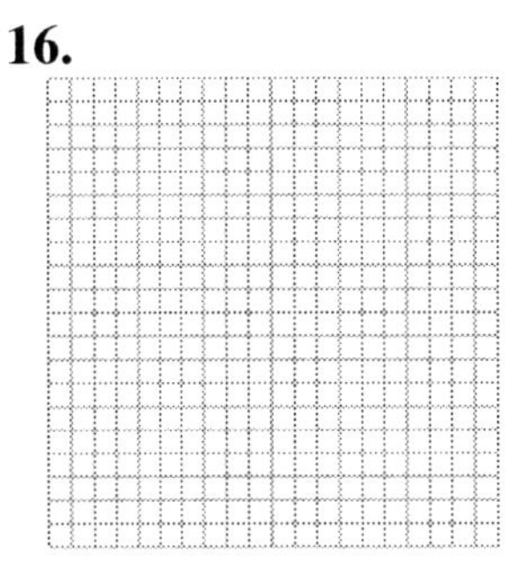

Name: Date:
Instructor: Section:

Chapter 3 SYSTEMS OF EQUATIONS

3.4 Systems of Three Equations in Three Unknowns

Learning Objectives

1. Determine whether an ordered triple is a solution of a system of three equations in three unknowns.
2. Solve systems of three equations in three unknowns.
3. Solve applied problems using a system of three linear equations in three unknowns.

Objective 1 Determine whether an ordered triple is a solution of a system of three equations in three unknowns.

Is the ordered triple a solution to the given system of equations?

1. $\left(\frac{1}{2}, \frac{2}{3}, 4\right)$ **1.**______________

$$\begin{aligned} 2x+3y+4z&=19 \\ 4x-3y-z&=-4 \\ 12x+12y-5z&=-6 \end{aligned}$$

2. $\left(-\frac{1}{4}, 0, \frac{2}{7}\right)$ **2.**______________

$$\begin{aligned} 4x-2y+7z&=1 \\ -4x+14y+21z&=7 \\ 20x-11y-7z&=-7 \end{aligned}$$

Name: Date:
Instructor: Section:

3. $(3, 5, 2)$

$$x + y + z = 10$$
$$x - 2y - z = -5$$
$$2x + 3y - z = 19$$

3.________________

Objective 2 Solve systems of three equations in three unknowns.

Solve.

4.

$$x + 3y - 2z = -3$$
$$-2x - 2y + 3z = -2$$
$$-3x + 4y - 5z = -17$$

4.________________

5.

$$3x + 4y + 5z = -15$$
$$4x + 5y - 6z = 20$$
$$5x - 6y + 7z = 55$$

5.________________

Name: Date:
Instructor: Section:

6.

$$\begin{aligned} x+y-z&=-9 \\ 3x+5y+6z&=-7 \\ 4y+7z&=-4 \end{aligned}$$

6.________________

7.

$$\begin{aligned} x+y+z&=1 \\ 2x+4y+2z&=-2 \\ -x+9y-3z&=-21 \end{aligned}$$

7.________________

8.

$$\begin{aligned} x+y+z&=1 \\ 2x+5y+2z&=17 \\ -x+8y-3z&=50 \end{aligned}$$

8.________________

Name: Date:
Instructor: Section:

9.

$$4x + 3y + z = 18$$
$$x - 3y + 2z = -12$$
$$11x - 2y + 3z = 18$$

9._______________

10.

$$7x - y + z = 17$$
$$3x + 2y - 3z = -22$$
$$x - 3y + 2z = 26$$

10._______________

11.

$$x + y - 4z = -16$$
$$2y + z = 3$$
$$z = 5$$

11._______________

Name: Date:
Instructor: Section:

Objective 3 Solve applied problems using a system of three linear equations in three unknowns.

Solve.

12. A basketball team sells tickets that cost \$10, \$20, or for VIP seats, \$30. The team has sold 492 tickets overall. It has sold 98 more \$20 tickets than \$10 tickets. The total sales are \$8680. How many tickets of each kind have been sold?

12.________________

13. In triangle *ABC*, the measure of angle *B* is $23°$ more than three times the measure of angle *A*. The measure of angle *C* is $52°$ more than the measure of angle *A*. Find the measure of each angle.

13.________________

Name: Date:
Instructor: Section:

14. An investment of $106,000 was made by a business club. The investment was split into three parts and lasted for one year. The first part of the investment earned 8% interest, the second 6%, and the third 9%. Total interest from the investments was $8100. The interest from the first investment was four times the interest from the second. Find the amounts of the three parts of the investment.

14.________________

Name: Date:
Instructor: Section:

Chapter 3 SYSTEMS OF EQUATIONS

3.5 Using Matrices to Solve Systems of Equations

Learning Objectives
1. Solve systems of two equations in two unknowns using matrices.
2. Solve systems of three equations in three unknowns using matrices.
3. Solve applied problems using matrices.

Key Terms
Use the most appropriate term or phrase from the given list to complete each statement in exercises 1-2.

elements **augmented** **upper triangular** **rows** **columns**

1. A 2×3 matrix has 2 ______________________ and 3 ______________________.

2. In an ______________________ matrix, the last column contains the constants of the system of equations corresponding to the matrix.

Objective 1 Solve systems of two equations in two unknowns using matrices.

Solve the system of equations using matrices.

1. $x+5y=34$
$4x+4y=8$

1.________________

2. $x-y=1$
$8x+7y=-22$

2.________________

Name: Date:
Instructor: Section:

3. $2x+9y=22$
$-6x+y=46$

3.________________

4. $9x-9y=-90$
$8x-y=-101$

4.________________

5. $7x+9y=25$
$3x-9y=-105$

5.________________

Objective 2 Solve systems of three equations in three unknowns using matrices.

Solve the system of equations using matrices.

6. $x+y+z=0$
$2x+5y+2z=15$
$-x+8y-3z=53$

6.________________

Name: Date:
Instructor: Section:

7. $x + y + z = -8$
$2x + 5y + 2z = -10$
$-x + 6y - 3z = 32$

7.________________

8. $3x + 3y + z = -21$
$x - 3y + 2z = 18$
$8x - 2y + 3z = -5$

8.________________

9. $6x - y + z = -22$
$2x + 2y - 3z = -31$
$x - 3y + 2z = 14$

9.________________

10. $5x - y + z = -14$
$2x + 2y - 3z = 19$
$x - 3y + 2z = -25$

10.________________

Name: Date:
Instructor: Section:

Objective 3 Solve applied problems using matrices.

Solve.

11. The College of the Sequoias Theater Department's spring musical had two performances: a matinee and an evening performance. Tickets to the matinee were $4 and tickets to the evening performance were $6.50. If a total of 623 people attended the two shows and paid a total of $3384.50, how many people attended the matinee and how many people attended the evening performance?

11.________________

12. To raise money for a field trip, students at a school sold t-shirts for $12 and sweatshirts for $30. The students sold $1728 worth of shirts. If the students sold twice as many t-shirts as sweatshirts, how many t-shirts did they sell?

12.________________

13. A motel clerk counts his $1 and $10 bills at the end of a day. He finds that he has a total of 52 bills having a combined monetary value of $151. Find the number of bills of each denomination that he has.

13.________________

Name: Date:
Instructor: Section:

Chapter 3 SYSTEMS OF EQUATIONS

3.6 Determinants and Cramer's Rule

Learning Objectives
1 Find the determinant of a 2×2 matrix.
2 Use Cramer's rule to solve systems of two linear equations in two unknowns.
3 Find the determinant of a 3×3 matrix.
4 Use Cramer's rule to solve systems of three linear equations in three unknowns.
5 Use Cramer's rule to solve applied problems.

Key Terms
Use the most appropriate term or phrase from the given list to complete each statement in exercises 1-2.

Cramer's rule **determinant** **square matrix** **minor**

1. A ______________________ has an equal number of rows and columns.

2. The ______________________ of an element in a matrix can be found by crossing out the row and column of the original determinant that contains that element.

Objective 1 Find the determinant of a 2 × 2 matrix.

Evaluate the determinant.

1. $\begin{vmatrix} 2 & 4 \\ -5 & 8 \end{vmatrix}$ **1.**________________

2. $\begin{vmatrix} -1 & 5 \\ 7 & 11 \end{vmatrix}$ **2.**________________

3. $\begin{vmatrix} 3 & -2 \\ -6 & 9 \end{vmatrix}$ **3.**________________

Name: Date:
Instructor: Section:

4. $\begin{vmatrix} 6 & -7 \\ -4 & 2 \end{vmatrix}$

4.________________

Objective 2 Use Cramer's rule to solve systems of two linear equations in two unknowns.

Solve the system of equations using Cramer's rule.

5. $\begin{aligned} x - 5y &= -9 \\ 3x + 2y &= 20 \end{aligned}$

5.________________

6. $\begin{aligned} 5x - 3y &= 5 \\ 10x + 9y &= 0 \end{aligned}$

6.________________

7. $\begin{aligned} 3x + 7y &= -12 \\ 9x - 5y &= 81 \end{aligned}$

7.________________

Name: Date:
Instructor: Section:

8. $$\begin{aligned}3x+2y&=5\\2x-4y&=22\end{aligned}$$

8.________________

9. $$\begin{aligned}6x+4y&=-34\\2x-4y&=-38\end{aligned}$$

9.________________

Objective 3 Find the determinant of a 3 × 3 matrix.

Evaluate the determinant.

10. $$\begin{vmatrix}2 & -3 & 0\\1 & 2 & -1\\1 & -2 & 4\end{vmatrix}$$

10.________________

11. $$\begin{vmatrix}-1 & 0 & 2\\-3 & 1 & 2\\-2 & 3 & 0\end{vmatrix}$$

11.________________

Name: Date:
Instructor: Section:

12. $\begin{vmatrix} 3 & 1 & 3 \\ -2 & 0 & 3 \\ -1 & 2 & 1 \end{vmatrix}$

12.________________

13. $\begin{vmatrix} 0 & 2 & -1 \\ 2 & -3 & 0 \\ 3 & 1 & 2 \end{vmatrix}$

13.________________

Objective 4 Use Cramer's rule to solve systems of three linear equations in three unknowns.

Solve the system of equations using Cramer's rule.

14.
$$\begin{aligned} x+y+z&=6 \\ 2x+5y+2z&=18 \\ -x+8y-3z&=6 \end{aligned}$$

14.________________

Name: Date:
Instructor: Section:

15.
$$\begin{aligned} x+y+z&=-6\\ 2x+5y+2z&=-3\\ -x+6y-3z&=33 \end{aligned}$$

15.________________

16.
$$\begin{aligned} 6x+3y+z&=-2\\ x-3y+2z&=1\\ 17x-2y+3z&=25 \end{aligned}$$

16.________________

17.
$$\begin{aligned} 6x-y+z&=34\\ 3x+2y-3z&=19\\ x-3y+2z&=-3 \end{aligned}$$

17.________________

18.
$$\begin{aligned} x+y-4z&=6\\ -3y+z&=9\\ z&=-3 \end{aligned}$$

18.________________

Name: Date:
Instructor: Section:

Objective 5 Use Cramer's rule to solve applied problems.

Solve using Cramer's rule.

19. An automobile dealer had 219 cars and trucks in stock during the month. He must pay an inventory fee of \$5 per car and \$6 per truck. If he paid \$1231 for inventory fees, how many cars and trucks did he have during the month?

19.________________

20. The Jurassic Zoo charges \$8 for each adult admission and \$3 for each child. The total bill for the 160 people from a school trip was \$650. How many adults and how many children went to the zoo?

20.________________

21. An investment of \$68,000 was made by a business club. The investment was split into three parts and lasted for one year. The first part of the investment earned 8% interest, the second 6%, and the third 9%. Total interest from the investments was \$5400. The interest from the first investment was 4 times the interest from the second. Find the amounts of the three parts of the investment.

21.________________

Name: Date:
Instructor: Section:

Chapter 4 EXPONENTS AND POLYNOMIALS

4.1 Exponents

Learning Objectives
1 Use the properties of exponents.
2 Use two or more properties of exponents to simplify an expression.
3 Evaluate functions using the rules for exponents.

Key Terms

Objective 1 Use the properties of exponents.

Simplify. (Assume all variables are nonzero.)

1. $x^{5y} \cdot x^{y}$ **1.**______________

2. $x^{3a} \cdot x^{4a}$ **2.**______________

3. $-6x^{3} \cdot 2x^{7}$ **3.**______________

4. $\left(x^{5}\right)^{3}$ **4.**______________

5. $\dfrac{g^{12}}{g^{12}}$ **5.**______________

6. $\dfrac{y^{32}}{y^{32}}$ **6.**______________

Name: Date:
Instructor: Section:

7. $\dfrac{x^{7n}}{x^{4n}}$ 7.________________

8. $\dfrac{x^{13n}}{x^{5n}}$ 8.________________

9. $\dfrac{(a-b)^{10}}{(a-b)^{7}}$ 9.________________

10. $\dfrac{4}{z^{-11}}$ 10.________________

11. $\dfrac{10}{x^{-5}}$ 11.________________

12. $\dfrac{16x^{6}y^{8}z^{3}}{2x^{5}y^{4}z}$ 12.________________

13. $-4x^{0}$ 13.________________

14. $6x^{2}y^{0}z^{3}$ 14.________________

15. $\left(\dfrac{6}{x}\right)^{3}$ 15.________________

Name: Date:
Instructor: Section:

Objective 2 Use two or more properties of exponents to simplify an expression.

Simplify. (Assume all variables are nonzero.)

16. $\left(x^3\right)^4\left(x^2\right)^5$ **16.**______________

17. $\left(3x^4\right)^3\left(4x^5\right)^2$ **17.**______________

18. $\left(2a^7\right)^5\left(-3a^6\right)^2$ **18.**______________

19. $\left(\dfrac{2s^{12}t^{15}}{3x^{10}y}\right)^2$ **19.**______________

20. $\left(\dfrac{a^{11}b^{21}}{c^6d^{30}}\right)^7$ **20.**______________

Find the missing exponent.

21. $\left(x^4\right)^? = x^{36}$ **21.**______________

22. $\left(b^?\right)^7 = b^{63}$ **22.**______________

23. $\left(x^3y^{13}\right)^?\left(x^7y^8\right)^5 = x^{47}y^{92}$ **23.**______________

Name: Date:
Instructor: Section:

24. $(a^5b^4)^6(a^3b^8)^? = a^{54}b^{88}$ **24.**________________

25. $\dfrac{t^{230}}{t^?} = 1$ **25.**________________

26. $\dfrac{a^?b^?}{a^8b^{21}} = a^{20}b^{49}$ **26.**________________

27. $\left(\dfrac{m^?}{n^?}\right)^6 = \dfrac{m^{18}}{n^{48}}$ **27.**________________

28. $\left(\dfrac{s^?}{t^?}\right)^{13} = \dfrac{s^{91}}{t^{143}}$ **28.**________________

Objective 3 Evaluate functions using the rules for exponents.

Evaluate the given function.

29. $f(x) = 3x^5$, $f(-5)$ **29.**________________

30. $f(x) = 7x^4$, $f(10)$ **30.**________________

Name: Date:
Instructor: Section:

Chapter 4 EXPONENTS AND POLYNOMIALS

4.2 Negative Exponents; Scientific Notation

Learning Objectives

1 Understand negative exponents.
2 Use the rules of exponents to simplify expressions containing negative exponents.
3 Convert numbers from standard notation to scientific notation.
4 Convert numbers from scientific notation to standard notation.
5 Perform arithmetic operations using numbers in scientific notation.
6 Use scientific notation to solve applied problems.

Objective 1 Understand negative exponents.

Rewrite the expression without using negative exponents. (Assume all variables represent nonzero real numbers.)

1. x^{-8} 1.________

2. -2^{-4} 2.________

3. $\dfrac{m^{-3}}{n^{-3}}$ 3.________

4. $\dfrac{a^{-5}}{b^{8}}$ 4.________

Objective 2 Use the rules of exponents to simplify expressions containing negative exponents.

Simplify the expression. Write the result without using negative exponents. (Assume all variables represent nonzero real numbers.)

5. $\left(x^{3}\right)^{-7}$ 5.________

Name: Date:
Instructor: Section:

6. $\left(-5x^{-3}y^{2}z^{-6}\right)^{3}$

6.________________

7. $\left(\dfrac{-3a^{5}b^{-11}}{c^{-4}d^{-7}}\right)^{2}$

7.________________

8. $\dfrac{x^{8}}{x^{-12}}$

8.________________

9. $\left(x^{2}y^{-7}\right)^{-5}$

9.________________

10. $\dfrac{m^{-7}}{m^{-18}}$

10.________________

11. $\left(\dfrac{x^{3}y^{4}}{2z^{6}}\right)^{6}$

11.________________

12. $\left(\dfrac{x^{-3}y^{-5}}{-4z^{2}w^{-9}}\right)^{3}$

12.________________

Name: Date:
Instructor: Section:

13. $\frac{x^8 \cdot x^{-5}}{x^{-32}}$

13._______________

Objective 3 Convert numbers from standard notation to scientific notation.

Convert the given number to scientific notation.

14. 275,000

14._______________

15. 92,300,000

15._______________

16. 0.00021

16._______________

17. 0.000036

17._______________

Objective 4 Convert numbers from scientific notation to standard notation.

Convert the given number to standard notation.

18. 4.561×10^{-3}

18._______________

19. 3.32×10^{-6}

19._______________

Name: Date:
Instructor: Section:

20. 2.6×10^9

20. ____________

21. 8.0×10^{11}

21. ____________

Objective 5 Perform arithmetic operations using numbers in scientific notation.

Perform the following calculations. Express your answer using scientific notation.

22. $\left(6.3 \times 10^4\right)\left(5.0 \times 10^{-10}\right)$

22. ____________

23. $\left(4.5 \times 10^5\right) + \left(3.21 \times 10^4\right)$

23. ____________

24. $\left(7.09 \times 10^6\right) - \left(5.15 \times 10^4\right)$

24. ____________

25. $\left(7.82 \times 10^6\right) \div \left(2.3 \times 10^{-9}\right)$

25. ____________

Name: Date:
Instructor: Section:

26. $\left(1.45\times10^{15}\right)\left(8.2\times10^{-4}\right)$

26.________________

Objective 6 Use scientific notation to solve applied problems.

Solve the following problems. Express your answer using scientific notation.

27. The mass of a hydrogen atom is 1.66×10^{-24} grams. What is the mass of 22,500,000,000,000,000 hydrogen atoms?

27.________________

28. The mass of a hydrogen atom is 1.66×10^{-24} grams. If the mass of a star is 8.3×10^{33} grams, how many hydrogen atoms are there in the star?

28.________________

29. The speed of light is 1.86×10^{5} miles per second. How far can light from the sun travel in 20 minutes?

29.________________

Name: Date:
Instructor: Section:

30. The speed of light is 1.86×10^5 miles per second.
a. How far could light from the sun travel in 12 minutes (720 seconds)?
b. How long would it take light from the sun to travel 9,300,000,000,000 miles?

30a.________________

b.________________

Name: Date:
Instructor: Section:

Chapter 4 EXPONENTS AND POLYNOMIALS

4.3 Polynomials; Addition, Subtraction, and Multiplication of Polynomials

Learning Objectives
1 Identify polynomials and understand the vocabulary used to describe them.
2 Evaluate polynomials.
3 Add and subtract polynomials.
4 Multiply polynomials.
5 Find special products.
6 Understand polynomials in several variables.

Objective 1 Identify polynomials and understand the vocabulary used to describe them.

List the degree of each term in the given polynomial.

1. $x^3 + 3x^5 - 9x^2 + x$ **1.**______________

2. $6x^7 - 9x^5 - 12x + 37$ **2.**______________

3. $5x^5 - 3x^4 + 2x^3 + x^2 - 15$ **3.**______________

List the coefficient of each term in the given polynomial

4. $x^5 + 9x^2 - x$ **4.**______________

5. $-x^4 + 8x^2 + 11x$ **5.**______________

Identify the given polynomial as a monomial, binomial, trinomial, or none of these.

6. $4x^4 - 9$ **6.**______________

7. $60x$ **7.**______________

8. $x^2 - 12x + 36$ **8.**______________

Name: Date:
Instructor: Section:

Objective 2 Evaluate polynomials.

Evaluate the polynomial for the given value of the variable.

9. $x^2 - 8x - 20$ for $x = 10$ **9.**____________

10. $x^3 - 7x^2 + 8x - 45$ for $x = 6$ **10.**____________

Evaluate the given polynomial function.

11. $f(x) = x^5 - 4x^3 - x - 13$, $f(4)$ **11.**____________

12. $f(x) = 2x^6 - 5x^5 - 13x^4 - 16x^2$, $f(3)$ **12.**____________

13. $f(x) = x^3 + 2x^2 - 35x - 56$, $f(-4)$ **13.**____________

Objective 3 Add and subtract polynomials.

Add or subtract.

14. $(4x^2 - 3x + 11) + (8x^2 + 6x - 40)$ **14.**____________

15. $(-11b^2 + 3b + 7) - (7b^2 + 6b - 3)$ **15.**____________

Name: Date:
Instructor: Section:

16. $\left(\frac{5}{6}x^2 - \frac{7}{5}x + \frac{3}{8}\right) + \left(\frac{3}{4}x^2 + \frac{5}{2}x + \frac{3}{4}\right)$

16.________________

Find the missing polynomial.

17. $\left(x^7 - 2x^5 + 9x^3 - x^2 + 14x - 100\right) - ?$
$= x^7 - 3x^6 - 6x^5 + 7x^4 + 8x^3 - x^2 + 5x - 71$

17.________________

18. $? + \left(10x^3 + 15x^2 - x + 24\right) = 13x^3 + 9x^2 - 5x - 32$

18.________________

For the given functions $f(x)$ and $g(x)$, find $f(x) + g(x)$ and $f(x) - g(x)$.

19. $f(x) = 4x^5 - 6x^3 - 13x^2 + 35$,
$g(x) = 4x^5 + 2x^4 + 6x^3 - 13x$

19.________________

____

20. $f(x) = 3x^2 + 8x - 25$, $g(x) = -2x^2 - 14x + 35$

20.________________

____

Objective 4 Multiply polynomials.

Multiply.

21. $-9x^6 \cdot 7x^3$

21.________________

Name: Date:
Instructor: Section:

22. $-10x^8(-4x^7)$ **22.**________________

23. $(-6x^4)(3x^6)(9x^{11})$ **23.**________________

Find the missing monomial.

24. $7x^5y^6z^3 \cdot ? = 49x^8y^7z^3$ **24.**________________

25. $13x^8y^7z^9 \cdot ? = 104x^9y^9z^{13}$ **25.**________________

Multiply.

26. $8x^3(7x^5 - 9x^3 - 16x)$ **26.**________________

27. $x^9(x^2 - 7x - 30)$ **27.**________________

Find the missing factor.

28. $-5x^5(?) = 5x^9 - 20x^7 + 45x^5$ **28.**________________

29. $-3x^3(?) = -21x^8 + 15x^7 + 27x^6$ **29.**________________

Name: Date:
Instructor: Section:

Multiply.

30. $(x^2+7x)(3x-8)$

30.________________

31. $(x^2+4)(x^2+13)$

31.________________

32. $(x-3)(x-17)$

32.________________

Find the missing term.

33. $(?+1)(3x-4)=12x^2-13x-4$

33.________________

34. $(5x-4)(?-6)=35x^2-58x+24$

34.________________

Given the functions $f(x)$ and $g(x)$, find $f(x)\cdot g(x)$.

35. $f(x)=13x^7,\ g(x)=-11x^{15}$

35.________________

36. $f(x)=x^2+5x-30,\ g(x)=-8x^2-15x+6$

36.________________

37. $f(x)=x-9,\ g(x)=x-12$

37.________________

Name: Date:
Instructor: Section:

Objective 5 Find special products.

Find the special product using the appropriate formula.

38. $(x+2)(x-2)$ **38.**________________

39. $(x-10)(x+10)$ **39.**________________

40. $(x-5)^2$ **40.**________________

41. $(3x+2)^2$ **41.**________________

Objective 6 Understand polynomials in several variables.

For each polynomial, list the degree of each term as well as the degree of the polynomial.

42. $10x^4-12x^2y+15xy^2$ **42.**________________

43. $-x^6-5x^3y^5+8y^{10}$ **43.**________________

Add or subtract.

44. $(11a^2t-2at^2+4at)+(-9a^2t-6at^2+5at)$ **44.**________________

45. $(10rf-6r^2f+15rf^2)-(7rf^2-4rf-15r^2f)$ **45.**________________

Name: Date:
Instructor: Section:

Chapter 4 EXPONENTS AND POLYNOMIALS

4.4 An Introduction to Factoring; The Greatest Common Factor; Factoring by Grouping

Learning Objectives
1 Find the greatest common factor (GCF) of two or more integers.
2 Find the GCF of two or more variable terms.
3 Factor the GCF out of each term of a polynomial.
4 Factor a common binomial factor from a polynomial.
5 Factor a polynomial by grouping.

Key Terms

Use the most appropriate term or phrase from the given list to complete each statement in exercises 1-3.

sum **term** **factor** **binomial** **monomial** **product**

1. For a variable factor to be included in the GCF of two or more variable terms, it must be a factor of each ____________________.

2. To factor a polynomial with four terms by grouping, factor a common factor out of the first two terms and another common factor out of the last two terms, then see if these two groups share a common ____________________ factor.

3. A polynomial has been factored when it is represented as the ____________________ of two or more polynomials.

Objective 1 Find the greatest common factor (GCF) of two or more integers.

Find the GCF.

1. 70, 98 **1.**______________

2. 28, 144 **2.**______________

Name: Date:
Instructor: Section:

Objective 2 Find the GCF of two or more variable terms.

Find the GCF.

3. $m^{11}n^5, m^7n^6$ **3.**________________

4. $x^3y^5z^7, x^6y^2z^3, x^9y^5z$ **4.**________________

5. $12x^6, 8x^5, 20x^4$ **5.**________________

6. $21a^2, 35a, 49a^3$ **6.**________________

Objective 3 Factor the GCF out of each term of a polynomial.

Factor the GCF out of the given expression.

7. $21x^8y^3+7x^4y^4$ **7.**________________

8. $6x-30x^2$ **8.**________________

9. $14x^5-2x$ **9.**________________

10. $15y^8-12y^5+4y^2$ **10.**________________

11. $21a+35a^2-28a^3$ **11.**________________

Name: Date:
Instructor: Section:

Objective 4 Factor a common binomial factor from a polynomial.

Factor the GCF out of the given expression.

12. $12x(9x+11)+(9x+11)$ **12.**________________

13. $x(6x+7)-(6x+7)$ **13.**________________

14. $6x(x^2+7)-5(x^2+7)$ **14.**________________

15. $7x(3x^2-5x)+9(3x^2-5x)$ **15.**________________

Objective 5 Factor a polynomial by grouping.

Factor by grouping.

16. $x^2-10x+3x-30$ **16.**________________

17. $x^2+5x-8x-40$ **17.**________________

18. $x^2-10x-3x+30$ **18.**________________

Name: Date:
Instructor: Section:

19. $2x^2+26x-3x-39$ **19.**________________

20. $9x^2-63x-8x+56$ **20.**________________

21. $6x^3+15x^2+10x+25$ **21.**________________

22. $2x^2-6x+5x-15$ **22.**________________

23. $12x^2-32x-21x+56$ **23.**________________

24. $x^3+12x^2-4x-48$ **24.**________________

25. $x^3+7x^2-4x-28$ **25.**________________

Name: Date:
Instructor: Section:

Chapter 4 EXPONENTS AND POLYNOMIALS

4.5 Factoring Trinomials of Degree 2

Learning Objectives
1 Factor a trinomial of the form $x^2 + bx + c$.
2 Factor a trinomial by first factoring out a common factor.
3 Factor a trinomial of the form $ax^2 + bx + c$ ($a \neq 1$) by grouping.
4 Factor a trinomial of the form $ax^2 + bx + c$ ($a \neq 1$) by trial and error.
5 Factor a perfect square trinomial.
6 Factor a trinomial in several variables.

Key Terms
Use the most appropriate term or phrase from the given list to complete each statement in exercises 1-2.

trinomial **perfect** **prime** **binomial** **factor**

1. A trinomial that cannot be factored is said to be ___________________.

2. If the factored form of a trinomial is the square of a binomial, the trinomial is called a ___________________ square trinomial.

Objective 1 Factor a trinomial of the form $x^2 + bx + c$.

Factor completely. If the polynomial cannot be factored, write "prime."

1. $x^2 + 18x + 72$ **1.**________________

2. $x^2 + 13x + 30$ **2.**________________

3. $x^2 - 13x + 42$ **3.**________________

4. $x^2 - 13x + 12$ **4.**________________

Name: Date:
Instructor: Section:

Factor completely. If the polynomial cannot be factored, write "prime."

5. $x^2+10x-75$ **5.**________________

6. $x^2+21x-72$ **6.**________________

7. $x^2+x-132$ **7.**________________

Objective 2 Factor a trinomial by first factoring out a common factor.

Factor completely. If the polynomial cannot be factored, write "prime."

8. $-2x^5+12x^4+80x^3$ **8.**________________

9. $7x^{12}-14x^{11}-245x^{10}$ **9.**________________

10. $4x^3-20x^2+52x$ **10.**________________

11. $-x^6-25x^5-46x^4$ **11.**________________

Name: Date:
Instructor: Section:

12. $5x^9 + 70x^8 + 225x^7$

12.________________

13. $9x^6 + 36x^5 + 36x^4$

13.________________

14. $100x^{10} - 1000x^9 + 2100x^8$

14.________________

15. $2x^2 - 14x - 48$

15.________________

Objective 3 Factor a trinomial of the form $ax^2 + bx + c$ ($a \neq 1$) by grouping.
Objective 4 Factor a trinomial of the form $ax^2 + bx + c$ ($a \neq 1$) by trial and error.

Factor completely. If the polynomial cannot be factored, write "prime."

16. $12x^2 - 80x - 75$

16.________________

17. $12x^2 - 53x + 20$

17.________________

Name: Date:
Instructor: Section:

18. $6x^2 + 3x - 135$

18.________________

19. $2x^2 - 4x - 30$

19.________________

20. $-2x^2 - 11x + 40$

20.________________

21. $12x^2 - 11x - 56$

21.________________

22. $2x^2 + 12x + 15$

22.________________

23. $12x^2 + x - 63$

23.________________

24. $x^2 + 6x - 55$

24.________________

Name: Date:
Instructor: Section:

25. $x^2 - 3x + 28$

25.________________

Objective 5 Factor a perfect square trinomial.

Factor completely. If the polynomial cannot be factored, write "prime."

26. $x^2 + 10x + 25$

26.________________

27. $x^2 - 6x + 9$

27.________________

28. $x^2 - 18x + 81$

28.________________

29. $x^2 + 8x + 16$

29.________________

Objective 6 Factor a trinomial in several variables.

Factor completely. If the polynomial cannot be factored, write "prime."

30. $x^2 - 5xy + 4y^2$

30.________________

31. $x^2 - 23xy + 60y^2$

31.________________

Name: Date:
Instructor: Section:

32. $x^2y^2 + 21xy + 38$

32.________________

33. $x^2y^2 - 7xy - 30$

33.________________

Name: Date:
Instructor: Section:

Chapter 4 EXPONENTS AND POLYNOMIALS

4.6 Factoring Special Binomials

Learning Objectives
1 Factor a difference of squares.
2 Factor the factors of a difference of squares.
3 Factor a difference of cubes.
4 Factor a sum of cubes.

Objective 1 Factor a difference of squares.

Factor completely. If the polynomial cannot be factored, write "prime."

1. $36 - x^2$ **1.**_______________

2. $-x^2 - 49$ **2.**_______________

3. $-12x^2 - 48y^2$ **3.**_______________

4. $x^2 - 81$ **4.**_______________

5. $10x^2 - 810$ **5.**_______________

6. $x^2 + 49$ **6.**_______________

7. $x^8 - 25$ **7.**_______________

Name: Date:
Instructor: Section:

8. $x^{12} - 4$ **8.**________________

9. $100 - x^4$ **9.**________________

10. $x^4 - 64$ **10.**________________

Objective 2 Factor the factors of a difference of squares.

Factor completely. If the polynomial cannot be factored, write "prime."

11. $x^8 - 1$ **11.**________________

12. $x^4 - 10{,}000$ **12.**________________

13. $x^8 - y^8$ **13.**________________

Objective 3 Factor a difference of cubes.

Factor completely. If the polynomial cannot be factored, write "prime."

14. $x^6 - 64y^3$ **14.**________________

15. $a^{21} - 125b^{15}$ **15.**________________

Name: Date:
Instructor: Section:

16. $-3x^3+24y^6$ **16.**______________

17. $8x^3-125y^3$ **17.**______________

18. x^9-64 **18.**______________

Objective 4 Factor a sum of cubes.

Factor completely. If the polynomial cannot be factored, write "prime."

19. $x^2y^3+125x^2$ **19.**______________

20. $1+a^3$ **20.**______________

21. $729+z^3$ **21.**______________

22. $125a^9b^3+64c^{12}$ **22.**______________

23. $1000x^{30}+729y^6z^{21}$ **23.**______________

Name: Date:
Instructor: Section:

24. $-5x^3 - 625$ **24.**________________

25. $-3x^3 - 1029$ **25.**________________

26. $x^6 + 343$ **26.**________________

Name: Date:
Instructor: Section:

Chapter 4 EXPONENTS AND POLYNOMIALS

4.7 Factoring Polynomials: A General Strategy

Learning Objectives
1 Understand the strategy for factoring a general polynomial.

Objective 1 Understand the strategy for factoring a general polynomial.

Factor completely. If the polynomial cannot be factored, write "prime."

1. $8x^3 - 27y^3$ **1.**________

2. $9x^3 - 27x^2 - 25x + 75$ **2.**________

3. $9x^3 - 36x^2 - x + 4$ **3.**________

4. $x^6 - 729$ **4.**________

5. $x^2 + 9x + 20$ **5.**________

6. $4x^2 + 8x - 96$ **6.**________

7. $x^2 + 13x + 36$ **7.**________

Name: Date:
Instructor: Section:

8. $5x^2 + 19x - 30$ **8.**________________

9. $x^2 + 15x + 54$ **9.**________________

10. $4x^2 + 20x + 25$ **10.**________________

11. $9x^2 + 24x + 16$ **11.**________________

12. $x^2 - 18xy + 81y^2$ **12.**________________

13. $x^2 + 2x - 63$ **13.**________________

14. $x^2 + 2x - 24$ **14.**________________

15. $2x^2 - x - 15$ **15.**________________

16. $3x^2 + 8x + 5$ **16.**________________

Name: Date:
Instructor: Section:

17. $16x^3 - 64x^2 - 25x + 100$

17.________________

18. $9x^3 - 9x^2 - 25x + 25$

18.________________

19. $x^2 - 6x - 216$

19.________________

20. $2x^2 + 4x - 48$

20.________________

21. $4x^2 + 12x - 72$

21.________________

22. $20x^2 - 11x - 3$

22.________________

23. $x^6 + 125$

23.________________

24. $x^6 + 8$

24.________________

25. $x^2 - 4x - 396$

25.________________

Name: Date:
Instructor: Section:

26. $x^2 - 10x - 299$

26.________________

27. $x^2 + 16x + 64$

27.________________

28. $x^2 + 8x + 16$

28.________________

29. $12x^2 + 40x - 32$

29.________________

30. $9x^2 + 30x - 24$

30.________________

Name: Date:
Instructor: Section:

Chapter 4 EXPONENTS AND POLYNOMIALS

4.8 Solving Quadratic Equations By Factoring

Learning Objectives
1 Solve an equation by using the zero-factor property of real numbers.
2 Solve a quadratic equation by factoring.
3 Solve a quadratic equation that is not in standard form.
4 Solve a quadratic equation that has coefficients that are fractions.
5 Find a quadratic equation, given its solutions.
6 Solve applied problems using quadratic equations.

Key Terms

Use the most appropriate term or phrase from the given list to complete each statement in exercises 1-3.

quadratic **standard** **numbers** **factors** **zero** ***x***

1. The ____________________ form of a quadratic equation is $ax^2+bx+c=0$.

2. The zero-factor property states that if a product is equal to zero then one of the ____________________ must be equal to zero.

3. To solve a quadratic equation by factoring, write the equation in standard form, factor the expression, and set each factor equal to ____________________.

Objective 1 Solve an equation by using the zero-factor property of real numbers.

Solve.

1. $(x+4)(x-5)=0$ **1.**________________

2. $(x-1)(x-6)=0$ **2.**________________

3. $(x-2)(x+5)(4x-3)=0$ **3.**________________

4. $x(3x-2)(3x+2)=0$ **4.**________________

Name: Date:
Instructor: Section:

5. $(3x+7)(x+6)=0$ **5.**________________

Objective 2 Solve a quadratic equation by factoring.

Solve.

6. $x^2+8x=0$ **6.**________________

7. $x^2-25x+24=0$ **7.**________________

8. $x^2+12x+32=0$ **8.**________________

9. $x^2-9=0$ **9.**________________

10. $9x^2+9x-180=0$ **10.**________________

Name: Date:
Instructor: Section:

Objective 3 Solve a quadratic equation that is not in standard form.

Solve.

11. $x^2 + 4x = 32$ **11.**________________

12. $x^2 = 10x - 25$ **12.**________________

13. $x^2 - 4x = 3x + 30$ **13.**________________

14. $x^2 = 25$ **14.**________________

15. $x^2 - 13x + 15 = 3x - 49$ **15.**________________

Objective 4 Solve a quadratic equation with coefficients that are fractions.

Solve.

16. $\frac{1}{4}x^2 - \frac{31}{16}x - \frac{45}{16} = 0$ **16.**________________

Name: Date:
Instructor: Section:

17. $\frac{1}{2}x^2 - \frac{3}{4}x - 5 = 0$

17.________________

18. $\frac{1}{2}x^2 - \frac{9}{4}x - \frac{5}{4} = 0$

18.________________

19. $\frac{1}{5}x^2 - \frac{3}{25}x - \frac{2}{25} = 0$

19.________________

20. $\frac{1}{3}x^2 - \frac{4}{9}x - \frac{5}{3} = 0$

20.________________

Objective 5 Find a quadratic equation, given its solutions.

Find a quadratic equation with integer coefficients that has the given solution set.

21. $\{3, 8\}$

21.________________

22. $\{6, -2\}$

22.________________

Name: Date:
Instructor: Section:

23. $\{0, -3\}$

23.______________

24. $\left\{\frac{4}{3}, -\frac{7}{11}\right\}$

24.______________

25. $\left\{\frac{10}{3}\right\}$

25.______________

26. $\{0, -6\}$

26.______________

27. $\left\{-\frac{1}{5}, \frac{1}{5}\right\}$

27.______________

28. Use the fact that $x = -\frac{27}{5}$ is a solution to the equation $5x^2 + 37x + 54 = 0$ to find the other solution to the equation.

28.______________

29. Use the fact that $x = -\frac{10}{23}$ is a solution to the equation $23x^2 - 197x - 90 = 0$ to find the other solution to the equation.

29.______________

Name: Date:
Instructor: Section:

30. Use the fact that $x = \frac{19}{6}$ is a solution to the equation $24x^2 - 142x + 209 = 0$ to find the other solution to the equation.

30.________________

Objective 6 Solve applied problems using quadratic equations.

Solve.

31. The product of two consecutive odd positive integers is 54 more than 5 times the smaller integer. Find the two integers.

31.________________

32. The difference of two positive numbers is 5 and their product is 300. Find the two numbers.

32.________________

33. If the perimeter of a rectangle is 36 inches and the area is 72 square inches, find the dimensions of the rectangle.

33.________________

Solve. Use the function $h(t) = -16t^2 + v_0 t + s$.

34. An object is launched upward from the ground with an initial velocity of 24 feet per second. How long will it take for the object to land on the ground?

34.________________

Name: Date:
Instructor: Section:

Chapter 5 RATIONAL EXPRESSIONS AND EQUATIONS

5.1 Rational Expressions and Functions

Learning Objectives
1 Evaluate rational expressions.
2 Find the values for which a rational expression is undefined.
3 Simplify rational expressions to lowest terms.
4 Identify factors in the numerator and denominator that are opposites.
5 Evaluate rational functions.
6 Find the domain of rational functions.

Key Terms

Use the most appropriate term or phrase from the given list to complete each statement in exercises 1-3.

rational	**numerator**	**denominator**	**undefined**	**defined**
terms	**expressions**	**function**	**opposites**	**nonzero**

1. The domain of a rational function includes all values for which the function is ______________________.

2. To simplify a rational expression, factor the ______________________ and ______________________ and divide out common factors.

3. A ______________________ expression is a quotient of two polynomials.

Objective 1 Evaluate rational expressions.

Evaluate the rational expression for the given value of the variable.

1. $\dfrac{10}{x^2+2x-3}$ for $x=3$ **1.**________________

2. $\dfrac{3x+10}{2x-5}$ for $x=5$ **2.**________________

Name: Date:
Instructor: Section:

3. $\frac{5x-22}{4x-19}$ for $x=-25$ 3.________________

4. $\frac{x-11}{x^2-6x+55}$ for $x=5$ 4.________________

5. $\frac{x^2-6x+15}{x^2+14x-7}$ for $x=8$ 5.________________

Objective 2 Find the values for which a rational expression is undefined.

Find all values of the variable for which the rational expression is undefined.

6. $\frac{8}{x+2}$ 6.________________

7. $\frac{2x}{7x+4}$ 7.________________

8. $\frac{x^2+6x+17}{x^2-10x+25}$ 8.________________

9. $\frac{x^2+3x-54}{x^2+5x-36}$ 9.________________

Name: Date:
Instructor: Section:

10. $\dfrac{x+8}{3x-5}$

10.________________

11. $\dfrac{x^2+6x-16}{x^2-64}$

11.________________

Objective 3 Simplify rational expressions to lowest terms.

Simplify the given rational expression. (Assume all denominators are nonzero.)

12. $\dfrac{x^2+13x+30}{x^2+2x-15}$

12.________________

13. $\dfrac{x^2+8x-33}{x^2-9}$

13.________________

14. $\dfrac{x^2+x-42}{x^2-8x+12}$

14.________________

15. $\dfrac{6x^5}{8x^7}$

15.________________

Name: Date:
Instructor: Section:

16. $\dfrac{x^2-7x+10}{x-2}$

16.________________

17. $\dfrac{x^2+7x-60}{x^2+20x+96}$

17.________________

Objective 4 Identify factors in the numerator and denominator that are opposites.

Determine whether the two given binomials are or are not opposites.

18. $x-7$ and $7-x$

18.________________

19. $x+1$ and $1+x$

19.________________

20. $4x-2$ and $2x-4$

20.________________

21. $7x-9$ and $-7x+9$

21.________________

Simplify the given rational expression. (Assume all denominators are nonzero.)

22. $\dfrac{(x+5)(7-x)}{(x+7)(x-7)(x-5)}$

22.________________

23. $\dfrac{32-8x}{x^2-13x+36}$

23.________________

Name: Date:
Instructor: Section:

Objective 5 Evaluate rational functions.

Evaluate the given rational function.

24. $r(x)=\dfrac{15}{x^2+8x-13}$, $r(-5)$

24.________________

25. $r(x)=\dfrac{x^2+2x-35}{x^2+11x+28}$, $r(-5)$

25.________________

Objective 6 Find the domain of rational functions.

Find the domain of the given rational function.

26. $r(x)=\dfrac{4}{(x+3)(3x-8)}$

26.________________

27. $r(x)=\dfrac{x^2+2x-3}{x^2-2x-15}$

27.________________

28. $r(x)=\dfrac{x^2+11x+24}{2x^2-13x-24}$

28.________________

29. $r(x)=\dfrac{x^2+6x-40}{x^2-10x+24}$

29.________________

Name: Date:
Instructor: Section:

Use the graph of a rational function r(x) to solve the problems that follow.

30. **30.a.**______________

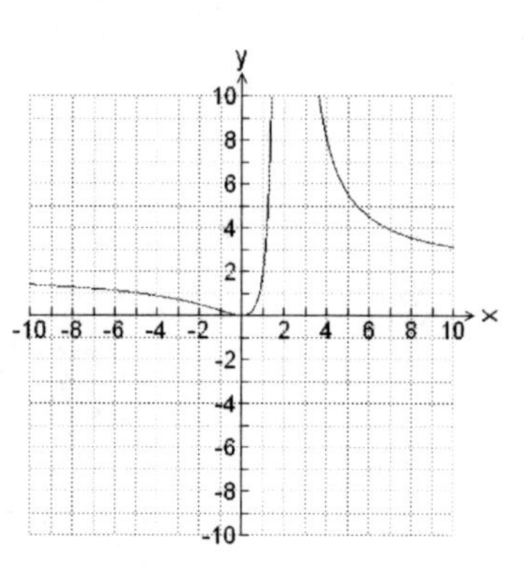

b.______________

a. Find $r(1)$.

b. Find all values x such that $r(x)=2$.

31. **31.a.**______________

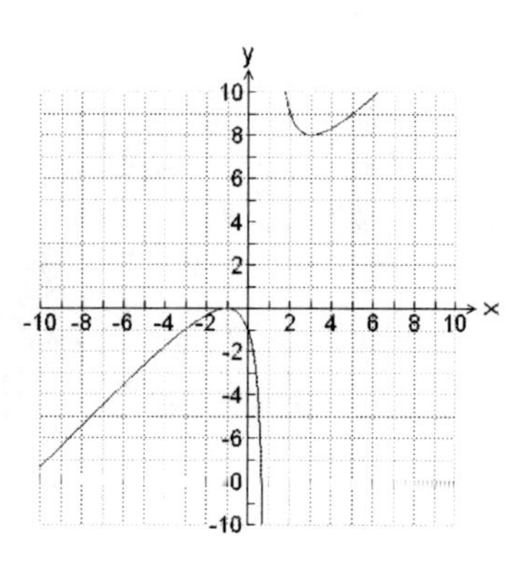

b.______________

a. Find $r(3)$.

b. Find all values x such that $r(x)=-1$.

Name: Date:
Instructor: Section:

Chapter 5 RATIONAL EXPRESSIONS AND EQUATIONS

5.2 Multiplication and Division of Rational Expressions

Learning Objectives
1 Multiply two rational expressions.
2 Multiply two rational functions.
3 Divide a rational expression by another rational expression.
4 Divide a rational function by another rational function.

Objective 1 Multiply two rational expressions.

Multiply.

1. $\dfrac{(x+3)}{(x-3)(x-5)}\cdot\dfrac{(x-5)(x+1)}{(x-2)(x+3)}$ **1.**______________

2. $\dfrac{x^2+15x+56}{x^2+2x-24}\cdot\dfrac{x^2-3x-4}{x^2+6x-16}$ **2.**______________

3. $\dfrac{x^2-2x-24}{3x^2+11x+10}\cdot\dfrac{3x^2-x-10}{x^2+10x+24}$ **3.**______________

4. $\dfrac{(x+4)(x-2)}{(x-2)(x-3)}\cdot\dfrac{(x+3)(x+8)}{(x+4)(x-5)}$ **4.**______________

5. $\dfrac{x+2}{(x-2)(x-4)}\cdot\dfrac{(x-4)(x+5)}{(x-3)(x+2)}$ **5.**______________

Name: Date:
Instructor: Section:

6. $\dfrac{x^2+16x+60}{x^2-13x+36}\cdot\dfrac{x^2-10x+24}{x^2+x-30}$

6.________________

Objective 2 Multiply two rational functions.

For the given functions f(x) and g(x), find f(x) · g(x).

7. $f(x)=\dfrac{x^2+3x-18}{x^2+4x-21},\ g(x)=\dfrac{x^2-4x-21}{x^2+9x+18}$

7.________________

8. $f(x)=\dfrac{x^2+2x-63}{x^2+3x-70},\ g(x)=\dfrac{x^2-4x-60}{x^2+15x+54}$

8.________________

Find the missing numerator and denominator.

9. $\dfrac{(x+8)(x-3)}{x(x+2)}\cdot\dfrac{?}{?}=-\dfrac{x+8}{x+2}$

9.________________

10. $\dfrac{x^2-14x+40}{x^2+2x-48}\cdot\dfrac{?}{?}=\dfrac{(x-10)(x-3)}{(x+8)(x-5)}$

10.________________

Name: Date:
Instructor: Section:

Objective 3 Divide a rational expression by another rational expression.

Divide.

11. $\frac{(3x+2)(x-2)}{(2x+1)(x-4)} \div \frac{3x+2}{x-4}$

11.________________

12. $\frac{x^2+4x-5}{x^2-49} \div \frac{x^2+3x-4}{x^2+5x-14}$

12.________________

13. $\frac{x^2+5x-6}{x^2+6x-7} \div \frac{x^2+8x+12}{x^2-5x-14}$

13.________________

14. $\frac{x^2+2x-48}{x^2-36} \div \frac{x^2+3x-54}{x^2+5x-6}$

14.________________

15. $\frac{(x+3)(x-4)}{x-9} \div \frac{(x-4)(x+1)}{(x-9)(x-2)}$

15.________________

16. $\frac{x^2-11x+18}{x^2+2x-3} \div \frac{x^2-5x-36}{x^2+10x+21}$

16.________________

Name: Date:
Instructor: Section:

Objective 4 Divide a rational function by another rational function.

For the given functions f(x) and g(x), find f(x) ÷ g(x).

17. $f(x) = \frac{x^2 + 6x - 7}{x^2 - 4x + 4}, \; g(x) = x + 7$

17.______________

18. $f(x) = \frac{5x^2 - x}{x^2 - 1}, \; g(x) = \frac{5x^2 + 39x - 8}{x^2 + 3x - 4}$

18.______________

Find the missing numerator and denominator.

19. $\frac{(2x+1)(x-5)}{(x+2)(x+6)} \div \frac{?}{?} = -\frac{2x+1}{x+6}$

19.______________

20. $\frac{x^2 - 12x + 35}{x^2 - x - 42} \div \frac{?}{?} = \frac{x+2}{x+8}$

20.______________

Name: Date:
Instructor: Section:

Chapter 5 RATIONAL EXPRESSIONS AND EQUATIONS

5.3 Addition and Subtraction of Rational Expressions

Learning Objectives
1. Add rational expressions with the same denominator.
2. Subtract rational expressions with the same denominator.
3. Add or subtract rational expressions with opposite denominators.
4. Find the least common denominator (LCD) of two or more rational expressions.
5. Add or subtract rational expressions with unlike denominators.

Objective 1 Add rational expressions with the same denominator.

Add.

1. $\frac{11}{2x-10}+\frac{15}{2x-10}$

1.______________

2. $\frac{x^2-4x+9}{x^2+9x+20}+\frac{x-37}{x^2+9x+20}$

2.______________

3. $\frac{x^2+10x+40}{x^2+12x+20}+\frac{6x+20}{x^2+12x+20}$

3.______________

4. $\frac{x(x+9)}{x^2+13x+36}+\frac{7(x+4)}{x^2+13x+36}$

4.______________

5. $\frac{x(x+8)}{x^2+5x-24}+\frac{20(x-3)}{x^2+5x-24}$

5.______________

Name: Date:
Instructor: Section:

6. $\dfrac{6x-10}{x^2-9x+20}+\dfrac{6-5x}{x^2-9x+20}$ 6.________________

7. $\dfrac{x^2-5x}{x^2+10x+25}+\dfrac{4x-30}{x^2+10x+25}$ 7.________________

8. $\dfrac{x+2}{x^2+7x+10}+\dfrac{3}{x^2+7x+10}$ 8.________________

For the given functions f(x) and g(x), find f(x) + g(x).

9. $f(x)=\dfrac{9}{x},\ g(x)=\dfrac{2}{x}$ 9.________________

10. $f(x)=\dfrac{x^2-6x-8}{x^2-12x+35},\ g(x)=\dfrac{5x-12}{x^2-12x+35}$ 10.________________

Objective 2 Subtract rational expressions with the same denominator.

Subtract.

11. $\dfrac{4}{x+10}-\dfrac{7}{x+10}$ 11.________________

12. $\dfrac{6}{x-6}-\dfrac{x}{x-6}$ 12.________________

Name: Date:
Instructor: Section:

13. $\frac{20x}{5x-1}-\frac{4}{5x-1}$

13.________________

14. $\frac{x+4}{x^2-81}-\frac{13}{x^2-81}$

14.________________

15. $\frac{2x^2+4}{x^2-11x+10}-\frac{x^2-23x+28}{x^2-11x+10}$

15.________________

16. $\frac{(x-10)(x+5)}{x^2-2x-8}-\frac{25(x-7)}{x^2-2x-8}$

16.________________

For the given functions f(x) and g(x), find f(x) – g(x).

17. $f(x)=\frac{16}{x+55}, g(x)=\frac{25}{x+55}$

17.________________

18. $f(x)=\frac{13}{x+42}, g(x)=\frac{36}{x+42}$

18.________________

Objective 3 Add or subtract rational expressions with opposite denominators.

Add or subtract.

19. $\frac{5x+7}{x-5}+\frac{3x}{5-x}$

19.________________

Name: Date:
Instructor: Section:

20. $\dfrac{x^2+9x}{x-1}-\dfrac{3x-13}{1-x}$

20.________________

21. $\dfrac{8x+1}{x-2}+\dfrac{5x}{2-x}$

21.________________

22. $\dfrac{x^2-2x-15}{5x-40}+\dfrac{3x+9}{40-5x}$

22.________________

23. $\dfrac{x^2-3x-25}{4x-36}+\dfrac{2x+11}{36-4x}$

23.________________

Objective 4 Find the least common denominator (LCD) of two or more rational expressions.

Find the LCD of the given rational expressions.

24. $\dfrac{-5}{2x}, \dfrac{10}{5x}$

24.________________

25. $\dfrac{5x}{x^2-4}, \dfrac{x}{x^2+6x-16}$

25.________________

26. $\dfrac{7x}{x^2-9}, \dfrac{x}{x^2+5x-24}$

26.________________

27. $\dfrac{x+3}{x^2+8x+16}, \dfrac{x-4}{x^2+7x+12}$

27.________________

Name: Date:
Instructor: Section:

28. $\dfrac{x+7}{x^2+16x+64}, \dfrac{x-8}{x^2+15x+56}$

28.________________

Objective 5 Add or subtract rational expressions with unlike denominators.

Add or subtract.

29. $\dfrac{4x}{3}+\dfrac{x}{5}$

29.________________

30. $\dfrac{5}{7x}-\dfrac{5}{8x}$

30.________________

31. $\dfrac{6}{sr^2}+\dfrac{5}{s^2r}$

31.________________

32. $\dfrac{2}{x-2}+\dfrac{2}{x+2}$

32.________________

33. $\dfrac{6}{x+8}-\dfrac{3}{x-8}$

33.________________

Name: Date:
Instructor: Section:

34. $\dfrac{2}{(x+2)(x+1)}+\dfrac{4}{(x+1)(x-1)}$

34.________________

35. $\dfrac{5}{x^2+8x+16}+\dfrac{9}{x^2-16}$

35.________________

36. $\dfrac{x}{x^2-1}-\dfrac{5}{x^2-2x+1}$

36.________________

37. $\dfrac{5x^2+11x+6}{3x-4}+\dfrac{2x^2-3x+30}{4-3x}$

37.________________

38. $\dfrac{x}{x^2-36}+\dfrac{7}{x^2-2x-48}$

38.________________

39. $\dfrac{x+2}{x^2+7x+12}-\dfrac{7}{x^2-x-12}$

39.________________

Name: Date:
Instructor: Section:

Chapter 5 RATIONAL EXPRESSIONS AND EQUATIONS

5.4 Complex Fractions

Learning Objectives
1 Simplify complex numerical fractions.
2 Simplify complex fractions containing variables.

Objective 1 Simplify complex numerical fractions.

Simplify the complex fraction.

1. $\dfrac{\frac{5}{9}+\frac{5}{6}}{4}$

1.________________

2. $\dfrac{6}{\frac{7}{10}-\frac{1}{8}}$

2.________________

3. $\dfrac{\frac{2}{3}+\frac{3}{4}}{\frac{11}{12}-\frac{5}{8}}$

3.________________

4. $\dfrac{4+\frac{1}{4}}{4-\frac{1}{5}}$

4.________________

Name: Date:
Instructor: Section:

Objective 2 Simplify complex fractions containing variables.

Simplify the complex fraction.

5. $\dfrac{1+\dfrac{19}{x}+\dfrac{90}{x^2}}{1+\dfrac{2}{x}-\dfrac{80}{x^2}}$

5.________________

6. $\dfrac{1+\dfrac{4}{x}-\dfrac{5}{x^2}}{1-\dfrac{10}{x}+\dfrac{9}{x^2}}$

6.________________

7. $\dfrac{x-\dfrac{1}{4}}{x+\dfrac{3}{8}}$

7.________________

8. $\dfrac{\dfrac{4}{x+4}-\dfrac{9}{x-1}}{\dfrac{x+8}{x-1}}$

8.________________

9. $\dfrac{1-\dfrac{4}{x^2}}{1+\dfrac{2}{\text{x}}}$

9.________________

Name: Date:
Instructor: Section:

Simplify the given rational expression, using the techniques developed in Sections 7.1 through 7.5.

10. $\dfrac{x^2+13x+22}{x^2-5x-14}$

10.________________

11. $\dfrac{x^2-5x-14}{x^2-9x+14}$

11.________________

12. $\dfrac{x^2+3x+7}{x^2+4x-12}+\dfrac{5x-27}{x^2+4x-12}$

12.________________

13. $\dfrac{x^2+14x+40}{x^2+8x+16}\cdot\dfrac{x^2-16}{x^2+5x-50}$

13.________________

14. $\dfrac{x^2-5x}{x^2+3x-4}\div\dfrac{x^2-14x+45}{x^2-9x+8}$

14.________________

15. $\dfrac{35}{8x-14}+\dfrac{20x}{14-8x}$

15.________________

16. $\dfrac{x^2-11x+52}{x-9}-\dfrac{9x-47}{x-9}$

16.________________

Name: Date:
Instructor: Section:

17. $\dfrac{x^2-x-6}{x^2-12x+27}\cdot\dfrac{x^2-16x+63}{x^2+4x+4}$

17.________________

18. $\dfrac{x^2-3x-28}{x^2-10x+16}\div\dfrac{x^2+12x+32}{x^2-12x+20}$

18.________________

19. $\dfrac{2x}{x^2-11x+30}-\dfrac{12}{x^2-11x+30}$

19.________________

20. $\dfrac{x^2-15x}{x-9}-\dfrac{54}{9-x}$

20.________________

21. $\dfrac{x+1}{x^2+2x-48}+\dfrac{2}{x^2+12x+32}$

21.________________

22. $\dfrac{x+8}{x^2+10x+24}-\dfrac{9}{x^2+3x-18}$

22.________________

23. $\dfrac{x^2-9x+14}{x^2+6x-16}$

23.________________

Name: Date:
Instructor: Section:

Chapter 5 RATIONAL EXPRESSIONS AND EQUATIONS

5.5 Rational Equations

Learning Objectives
1 Solve rational equations.
2 Solve literal equations.

Key Terms
Use the most appropriate term or phrase from the given list to complete each statement in exercises 1-3.

undefined **denominators** **variables** **equations** **expressions**

1. A literal equation is an equation containing two or more ____________________.

2. An extraneous solution of a rational equation causes one of the expressions to be ____________________.

3. When solving a rational equation, use the LCD to clear the ____________________ of the rational expressions.

Objective 1 Solve rational equations.

Solve.

1. $\frac{11x}{12}+8=\frac{3x}{10}$ **1.**____________________

2. $\frac{5x}{8}+x=\frac{13}{5}$ **2.**____________________

3. $\frac{4}{5x}+\frac{11}{3}=\frac{8}{15}$ **3.**____________________

Name: Date:
Instructor: Section:

4. $\dfrac{3}{2x}-13=\dfrac{1}{9x}$

4.________________

5. $2x+\dfrac{7}{x}=x+\dfrac{32}{x}$

5.________________

6. $x+10=\dfrac{56}{x}$

6.________________

7. $6+\dfrac{x+10}{3x+5}=\dfrac{5x-2}{3x+5}$

7.________________

8. $3+\dfrac{5x+7}{x-4}=\dfrac{4x+11}{x-4}$

8.________________

9. $\dfrac{x}{x-4}=\dfrac{x+5}{x+3}$

9.________________

10. $\dfrac{x-8}{x-2}=\dfrac{x+6}{x+10}$

10.________________

Name: Date:
Instructor: Section:

11. $\dfrac{2x-11}{x-1}=\dfrac{x-7}{x+3}$

11.________________

12. $\dfrac{3x+13}{2x-3}=\dfrac{x+3}{x-3}$

12.________________

13. $\dfrac{3}{x^2+11x+30}=\dfrac{4}{x^2-2x-48}$

13.________________

14. $\dfrac{2x+27}{x^2-9x+14}-\dfrac{5}{x-7}=\dfrac{x-4}{x-2}$

14.________________

15. $\dfrac{35-6x}{x^2+17x+70}-\dfrac{x+5}{x+10}=\dfrac{4}{x+7}$

15.________________

16. $\dfrac{x+8}{x+9}+1=\dfrac{4-2x}{x^2+16x+63}$

16.________________

17. $\dfrac{x+1}{x-3}+\dfrac{x-2}{x^2-13x+30}=2$

17.________________

Name: Date:
Instructor: Section:

18. $\frac{x+5}{x^2-4x+3}+\frac{x+2}{x^2-5x+6}=\frac{x+22}{x^2-3x+2}$

18.________________

19. $\frac{x+10}{x^2-3x-10}+\frac{x-2}{x^2-4x-12}=\frac{x-4}{x^2-11x+30}$

19.________________

20. $8+\frac{6}{x-4}=\frac{x-12}{x-4}$

20.________________

Objective 2 Solve literal equations containing rational expressions.

Solve for the specified variable.

21. $\frac{1}{a}+\frac{1}{b}=\frac{1}{c}$ for c

21.________________

22. $\frac{C}{P}=T$ for P

22.________________

23. $D=\frac{5G}{M+5s}$ for M

23.________________

Name: Date:
Instructor: Section:

Chapter 5 RATIONAL EXPRESSIONS AND EQUATIONS

5.6 Applications of Rational Equations

Learning Objectives
1 Solve applied problems involving the reciprocal of a number.
2 Solve applied work-rate problems.
3 Solve applied uniform motion problems.
4 Solve variation problems.

Key Terms
Use the most appropriate term or phrase from the given list to complete each statement in exercises 1-3.

directly **inversely** **variation** **rate** **unknown**

1. When two or more people work together to perform a job, to determine the portion of the job completed by each person we must know the ________________ at which each person works.

2. Two quantities vary ________________ if an increase in one quantity produces a proportional decrease in the other quantity.

3. Two quantities vary ________________ if an increase in one quantity produces a proportional increase in the other quantity.

Objective 1 Solve applied problems involving the reciprocal of a number.

Solve.

1. The sum of the reciprocal of a number and $\frac{5}{9}$ is $\frac{13}{18}$. Find the number.

1.________________

2. One positive number is 14 less than another positive number. If the reciprocal of the smaller number is added to five times the reciprocal of the larger number, the sum is $\frac{1}{4}$. Find the two numbers.

2.________________

Name: Date:
Instructor: Section:

3. The difference of two positive numbers is 8. If the reciprocal of the smaller number is added to three times the reciprocal of the larger number, the sum is equal to $\frac{1}{2}$. Find the two numbers.

3.______________

4. The sum of two positive numbers is 15. If two times the reciprocal of the smaller number is added to the reciprocal of the larger number, the sum is equal to $\frac{1}{2}$. Find the two numbers.

4.______________

Objective 2 Solve applied work-rate problems.

Solve.

5. Steve takes 8 hours longer than Jan to wire a new house. Steve worked alone on wiring a new house for 4 hours before Jan began to help him. Once Jan began to help, the job took 2 hours to finish. How long would it take Steve to wire a new house by himself?

5.______________

Name: Date:
Instructor: Section:

6. A college has two printers available to print out class rosters for the first day of classes. The older printer takes 10 hours longer than the newer printer to print out a complete set of class rosters, so the dean of registration decides to use the newer printer. After 3 hours the newer printer breaks down, so the dean must switch to the older printer. It takes the older printer an additional 8 hours to complete the job. How long would it take the older printer to print out a complete set of class rosters?

6.________________

7. A bricklayer's apprentice takes 10 hours longer than the bricklayer to make a fireplace. The apprentice worked alone on a fireplace for 5 hours, then the bricklayer began to help. It took 2 more hours for the pair to finish the fireplace. How long would it take the apprentice to make a fireplace on his own?

7.

8. Two copy machines can print a set of training manuals in 12 hours. It would take the slower machine 18 hours longer than it would take the other machine to print the entire set of manuals. How long would it take the slower machine to print the entire set of manuals if it was working alone?

8.________________

Name: Date:
Instructor: Section:

9. Dina can paint a room in 30 minutes, while Scott takes 60 minutes to paint the same size room. If Dina and Scott work together, how long would it take them to paint a room?

9.________________

Objective 3 Solve applied uniform motion problems.

Solve.

10. An airplane flying to the west travels 100 miles per hour slower than an airplane flying to the east. If the westbound plane can fly 2000 miles in the same time that the eastbound plane can travel 2400 miles, find the airspeed of the eastbound airplane.

10.________________

11. A passenger train travels 50 kilometers per hour faster than a freight train does. If the passenger train can travel 300 kilometers in the same time that a freight train can travel 180 kilometers, find the speed of the passenger train.

11.________________

12. An airplane flies 48 miles with a 20 mile per hour tailwind, and then turns around and flies home into the 20 mile per hour wind. If the time for the round trip was 1 hour, find the speed of the airplane in calm air.

12.________________

Objective 4 Solve variation problems.

Solve.

13. The amount of sugar in a glass of milk varies directly as the amount of milk. If an 8-ounce serving of milk has 14 grams of sugar, how many grams of sugar are there in a 12-ounce glass of milk?

13.________________

14. The amount of fertilizer required for a lawn varies directly as the area of the lawn. If a lawn has an area of 1000 square feet, then it requires 8 pounds of fertilizer. How many pounds of fertilizer are needed for a lawn whose area is 3500 square feet?

14.________________

Name: Date:
Instructor: Section:

15. The time required to drive from one city to another varies inversely as the average speed of the car. If it takes 6 hours to make the drive at 50 miles per hour, how fast would you have to drive to arrive in 4 hours?

15._______________

16. The amount of time required to drain a swimming pool varies inversely as the rate at which the pump operates. If a pump's rate is 800 gallons per minute, it takes 50 minutes to drain the pool. How long would it take a pump that can pump 1000 gallons per minute to drain the pool?

16._______________

17. The distance that a free-falling object will fall varies directly as the square of the time that it is falling. An object that is falling for 4 seconds falls a distance of 256 feet. How far does an object fall if it is falling for 5 seconds?

17._______________

18. The distance required for a car to stop after applying the brakes varies directly as the square of the speed of the car. If it takes 96 feet for a car traveling at 40 miles per hour to stop, how far would it take for a car traveling 80 miles per hour to come to a stop?

18._______________

Name: Date:
Instructor: Section:

Chapter 5 RATIONAL EXPRESSIONS AND EQUATIONS

5.7 Division of Polynomials

Learning Objectives
1 Divide a monomial by a monomial.
2 Divide a polynomial by a monomial.
3 Divide a polynomial by a polynomial using long division.
4 Use placeholders when dividing a polynomial by a polynomial.

Objective 1 Divide a monomial by a monomial.

Divide. Assume all variables are nonzero.

1. $\dfrac{16x^5}{2x^6}$

1.________________

2. $\dfrac{-14x^4}{-7x^2}$

2.________________

3. $\dfrac{46r^{12}}{2r^6}$

3.________________

4. $\dfrac{15u^4v^3}{-3uv}$

4.________________

Find the missing monomial. Assume $x \neq 0$.

5. $\dfrac{?}{4x^2y^3} = 7x^2y$

5.________________

Name: Date:
Instructor: Section:

6. $\dfrac{32x^4y^6}{?} = 4y$

6.________________

Objective 2 Divide a polynomial by a monomial.

Divide. Assume all variables are nonzero.

7. $\dfrac{-14x^4 + 12x^3 - 8x^2}{-2x^2}$

7.________________

8. $\dfrac{33x^9 - 44x^7 + 55x^5}{11x^4}$

8.________________

9. $\dfrac{8x^6 - 40x^5 + 56x^4 + 72x^3}{8x^3}$

9.________________

10. $\dfrac{30x^7 - 15x^5 - 40x^4 + 5x^3}{-5x^3}$

10.________________

Find the missing dividend. Assume $x \neq 0$.

11. $\dfrac{?}{-4x^5} = 2x^3 - 5x - 11$

11.________________

Name: Date:
Instructor: Section:

12. $\dfrac{?}{8x^2} = x^4 - 3x^2 + 4x - 8 + \dfrac{2}{x}$

12.________________

Objective 3 Divide a polynomial by a polynomial using long division.

Divide using long division

13. $\dfrac{12x^2 - 11x - 35}{4x + 3}$

13.________________

14. $\dfrac{15x^2 - 16x - 391}{3x - 17}$

14.________________

15. $\dfrac{x^2 + 7x + 10}{x + 5}$

15.________________

16. $\dfrac{x^2 + 2x - 45}{x + 7}$

16.________________

Name: Date:
Instructor: Section:

Find the missing dividend.

17. $\dfrac{?}{x-6}=x-9+\dfrac{3}{x-6}$

17.________________

18. $\dfrac{?}{x+12}=x+11+\dfrac{7}{x+12}$

18.________________

Objective 4 Use placeholders when dividing a polynomial by a polynomial.

Divide using long division.

19. $\dfrac{x^3-9x+29}{x-6}$

19.________________

20. $\dfrac{a^3-a+2}{a-2}$

20.________________

21. $\dfrac{x^3+64}{x+4}$

21.________________

22. $\dfrac{x^3+216}{x+6}$

22.________________

Name: Date:
Instructor: Section:

Chapter 6 RADICAL EXPRESSIONS AND EQUATIONS

6.1 Square Roots; Radical Notation

Learning Objectives
1 Find the square root of a number.
2 Simplify the square root of a variable expression.
3 Approximate the square root of a number by using a calculator.
4 Find *n*th roots.
5 Multiply radical expressions.
6 Divide radical expressions.
7 Evaluate radical functions.
8 Find the domain of a radical function.

Key Terms

Use the most appropriate term or phrase from the given list to complete each statement in exercises 1-2.

principal **radical** **radicand** **real** **imaginary** **root**

1. The $\sqrt{\ }$ symbol is called a ______________________ sign.

2. If $a^n = b$ and a and b both have the same sign and n is an integer greater than 1, then a is the principal nth ______________________ of b.

Objective 1 Find the square root of a number.

Simplify the radical expression.

1. $\sqrt{36}$ **1.**________________

2. $\sqrt{289}$ **2.**________________

3. $\sqrt{\frac{25}{49}}$ **3.**________________

Name: Date:
Instructor: Section:

Objective 2 Simplify the square root of a variable expression.

Simplify the radical expression. Where appropriate, include absolute values.

4. $\sqrt{x^8}$ **4.**______________

5. $\sqrt{121x^{14}}$ **5.**______________

6. $\sqrt{x^{18}y^{26}}$ **6.**______________

Find the missing number or expression. Assume that all variables represent nonnegative real numbers.

7. $\sqrt{?} = m^8 n^9$ **7.**______________

8. $\sqrt{?} = x^{15} n^{13}$ **8.**______________

Objective 3 Approximate the square root of a number by using a calculator.

Approximate to the nearest thousandth, using a calculator.

9. $\sqrt{23,485}$ **9.**______________

10. $\sqrt{90,210}$ **10.**______________

11. $\sqrt{99}$ **11.**______________

Name: Date:
Instructor: Section:

12. $\sqrt{145}$

12.________________

Objective 4 Find *n*th roots.

Simplify the radical expression. Assume that all variables represent nonnegative real numbers.

13. $\sqrt[6]{x^{30}}$

13.________________

14. $\sqrt[8]{x^{56}}$

14.________________

15. $\sqrt[3]{216}$

15.________________

16. $\sqrt[4]{x^{20}}$

16.________________

17. $\sqrt[3]{-8x^6y^{15}z^3}$

17.________________

Objective 5 Multiply radical expressions.

Simplify.

18. $\sqrt[3]{9}\cdot\sqrt[3]{24}$

18.________________

Name: Date:
Instructor: Section:

19. $\sqrt{3} \cdot \sqrt{27}$ **19.**______________

20. $5\sqrt{6} \cdot \sqrt{294}$ **20.**______________

Find the missing number.

21. $\sqrt{16} \cdot \sqrt{?} = 32$ **21.**______________

22. $\sqrt{64} \cdot \sqrt{?} = 40$ **22.**______________

Objective 6 Divide radical expressions.

Simplify.

23. $\dfrac{\sqrt{63}}{\sqrt{7}}$ **23.**______________

24. $\dfrac{\sqrt{45}}{\sqrt{5}}$ **24.**______________

Name: Date:
Instructor: Section:

25. $\dfrac{\sqrt{250}}{\sqrt{10}}$

25.________________

Find the missing number.

26. $\dfrac{\sqrt{252}}{\sqrt{?}}=3$

26.________________

27. $\dfrac{\sqrt{864}}{\sqrt{?}}=4$

27.________________

Objective 7 Evaluate radical functions.

Evaluate the radical function. (Round to the nearest thousandth if necessary.)

28. $f(x)=\sqrt{x^2+11x+27}$, find $f(-9)$

28.________________

29. $f(x)=\sqrt{x^2+4x-22}$, find $f(-10)$

29.________________

30. $f(x)=\sqrt{5x-19}-16$, find $f(47)$

30.________________

Name: Date:
Instructor: Section:

Objective 8 Find the domain of a radical function.

Find the domain of the radical function. Express your answer in interval notation.

31. $f(x)=\sqrt{12-6x}$ **31.**______________

32. $f(x)=\sqrt{24-3x}-16$ **32.**______________

33. $f(x)=\sqrt{7x-56}$ **33.**______________

Name: Date:
Instructor: Section:

Chapter 6 RADICAL EXPRESSIONS AND EQUATIONS

6.2 Rational Exponents

Learning Objectives

1 Simplify expressions containing exponents of the form $1/n$.
2 Simplify expressions containing exponents of the form m/n.
3 Simplify expressions containing rational exponents.
4 Simplify expressions containing negative rational exponents.
5 Use rational exponents to simplify radical expressions.

Objective 1 Simplify expressions containing exponents of the form 1/*n*.

Rewrite each radical expression using rational exponents.

1. $\sqrt[3]{b}$ **1.**________________

2. $\sqrt[5]{s}$ **2.**________________

3. $\sqrt{7}$ **3.**________________

4. $\sqrt[7]{z}$ **4.**________________

Rewrite as a radical expression and simplify if possible. Assume that all variables represent nonnegative real numbers.

5. $\left(x^{21}y^{14}z^{49}\right)^{1/7}$ **5.**________________

6. $\left(a^{44}b^{80}c^{28}\right)^{1/4}$ **6.**________________

7. $\left(81a^{34}b^{22}c^{8}\right)^{1/2}$ **7.**________________

Name: Date:
Instructor: Section:

8. $\left(49x^{10}y^{10}z^{2}w^{200}\right)^{1/2}$ **8.**________________

9. $125^{1/3}$ **9.**________________

10. $\left(81x^{8}\right)^{1/4}$ **10.**________________

Objective 2 Simplify expressions containing exponents of the form *m/n*.

Rewrite each radical expression using rational exponents.

11. $\sqrt[4]{\left(2x^{2}y^{3}\right)^{3}}$ **11.**________________

12. $\sqrt[14]{\left(5x^{7}y^{3}z^{2}\right)^{5}}$ **12.**________________

13. $\sqrt{m^{9}}$ **13.**________________

14. $\sqrt[10]{x^{7}}$ **14.**________________

Rewrite as a radical expression and simplify if possible. Assume that all variables represent nonnegative real numbers.

15. $100{,}000{,}000^{5/8}$ **15.**________________

Name: Date:
Instructor: Section:

16. $\left(100{,}000x^{25}y^{35}z^{50}\right)^{3/5}$ **16.**________________

17. $\left(1{,}000{,}000x^{66}y^{36}z^{6}\right)^{5/6}$ **17.**________________

18. $(-8)^{7/3}$ **18.**________________

19. $16^{3/4}$ **19.**________________

20. $\left(9x^{8}\right)^{5/2}$ **20.**________________

21. $\left(8x^{21}y^{15}\right)^{7/3}$ **21.**________________

Objective 3 Simplify expressions containing rational exponents.

Simplify the expression. Assume that all variables represent nonnegative real numbers.

22. $x^{3/4}y^{7/12}\cdot x^{9/4}y^{11/12}$ **22.**________________

23. $x^{4/5}y^{2/3}\cdot x^{7/15}y^{5/6}$ **23.**________________

Name: Date:
Instructor: Section:

24. $x^{2/5} \cdot x^{1/10}$ **24.**________________

25. $\left(32x^{8/9}\right)^{3/5}$ **25.**________________

26. $\dfrac{x^{8/5}}{x^{3/2}}$ $(x \neq 0)$ **26.**________________

27. $\dfrac{x^{11/6}}{x^{7/12}}$ $(x \neq 0)$ **27.**________________

28. $\dfrac{x^{11/28}}{x^{1/7}}$ **28.**________________

29. $\dfrac{x^{11/6}}{x^{7/8}}$, $(x \neq 0)$ **29.**________________

30. $\left(8x^{1/6}y^{2/3}\right)^{0}$ $(x, y \neq 0)$ **30.**________________

31. $\left(14x^{7/5}y^{3/5}\right)^{0}$ $(x, y \neq 0)$ **31.**________________

Name: Date:
Instructor: Section:

32. $\left(x^{4/5}\right)^{9/2}$

32.________________

33. $\dfrac{16^{1/4}}{16^{1/2}}$

33.________________

Objective 4 Simplify expressions containing negative rational exponents.

Simplify the expression. Assume that all variables represent nonnegative real numbers.

34. $8^{-1/3}$

34.________________

35. $9^{-7/2}$

35.________________

36. $256^{-5/4}$

36.________________

37. $2^{-1/5} \cdot 2^{-4/5}$

37.________________

Objective 5 Use rational exponents to simplify radical expressions.

Simplify each expression. Assume all variables represent positive real numbers. Express your answer in radical notation.

38. $\sqrt[3]{m}\sqrt{m}$

38.________________

Name: Date:
Instructor: Section:

39. $\dfrac{\sqrt[5]{w}}{\sqrt[6]{w}}$

39.________________

40. $\dfrac{\sqrt[5]{y}}{\sqrt[10]{y}}$

40.________________

Name: Date:
Instructor: Section:

Chapter 6 RADICAL EXPRESSIONS AND EQUATIONS

6.3 Simplifying, Adding, and Subtracting Radical Expressions

Learning Objectives
1. Simplify radical expressions by using the product property.
2. Add or subtract radical expressions containing like radicals.
3. Simplify radical expressions before adding or subtracting.

Objective 1 Simplify radical expressions using the product property.

Simplify.

1. $\sqrt{45}$ **1.**________________

2. $\sqrt{90}$ **2.**________________

3. $\sqrt[3]{162}$ **3.**________________

4. $\sqrt[3]{750}$ **4.**________________

Simplify the radical expression. Assume that all variables represent nonnegative real numbers.

5. $\sqrt[3]{x^{40}y^{17}z^{45}}$ **5.**________________

6. $\sqrt[3]{xyz^{25}}$ **6.**________________

Name: Date:
Instructor: Section:

7. $\sqrt[3]{250x^{15}yz^{20}}$ 7.________

8. $\sqrt[3]{108x^{42}y^{26}z^{31}}$ 8.________

9. $\sqrt[3]{405a^6b^{10}c^2}$ 9.________

Objective 2 Add or subtract radical expressions containing like radicals.

Add or subtract. Assume that all variables represent nonnegative real numbers.

10. $15\sqrt[3]{7}-19\sqrt[3]{7}$ 10.________

11. $9\sqrt[4]{x^3}+11\sqrt[4]{x^3}$ 11.________

12. $13\sqrt{x}-5\sqrt{x}$ 12.________

13. $2\sqrt{22}+10\sqrt[3]{22}-15\sqrt{22}+3\sqrt[3]{22}$ 13.________

14. $\sqrt{3}+8\sqrt[3]{3}+7\sqrt[3]{3}-25\sqrt{3}$ 14.________

Name: Date:
Instructor: Section:

15. $5a\sqrt{n}-3\sqrt{m}+14a\sqrt{n}-8\sqrt{m}$ 15.________________

16. $6x\sqrt{x}+b\sqrt{y}+15x\sqrt{x}+9b\sqrt{y}$ 16.________________

Objective 3 Simplify radical expressions before adding or subtracting.

Add or subtract. Assume that all variables represent nonnegative real numbers.

17. $2a\sqrt{45}-\sqrt{5a^2}+11a\sqrt{180}$ 17.________________

18. $5\sqrt{44x^4}-3x\sqrt{99x^2}+8x^2\sqrt{176}$ 18.________________

19. $\left(8\sqrt{175}+9\sqrt{243}\right)-\left(5\sqrt{64}-7\sqrt{252}\right)$ 19.________________

20. $\left(5\sqrt{15}-12\sqrt{20}\right)-\left(8\sqrt{60}-14\sqrt{375}\right)$ 20.________________

Name: Date:
Instructor: Section:

21. $6\sqrt{27}+5\sqrt{50}-7\sqrt{20}+4\sqrt{300}$

21.________________

22. $7\sqrt{48}-16\sqrt{27}$

22.________________

23. $2\sqrt{63}+5\sqrt{147}+13\sqrt{243}-21\sqrt{252}$

23.________________

Find the missing radical expression.

24. $\left(10\sqrt{180}-12\sqrt{75}\right)+(?)=11\sqrt{3}-78\sqrt{5}$

24.________________

25. $\left(9\sqrt{6}-2\sqrt{10}\right)-(?)=5\sqrt{6}+8\sqrt{10}$

25.________________

Name: Date:
Instructor: Section:

Chapter 6 RADICAL EXPRESSIONS AND EQUATIONS

6.4 Multiplying and Dividing Radical Expressions

Learning Objectives
1 Multiply radical expressions.
2 Use the distributive property to multiply radical expressions.
3 Multiply radical expressions that have two or more terms.
4 Multiply radical expressions that are conjugates.
5 Rationalize a denominator that has one term.
6 Rationalize a denominator that has two terms.

Key Terms
Use the most appropriate term or phrase from the given list to complete each statement in exercises 1-3.

conjugate **rationalize** **numerator** **denominator** **opposite**

1. To simplify the expression $\frac{\sqrt{5}}{\sqrt{7}}$, multiply the numerator and denominator by the expression in the ________________.

2. The expressions $a + b$ and $a - b$ are called ________________ expressions.

3. To rationalize a denominator that has two terms with at least one term being a square root, multiply the numerator and denominator by the ________________ of the denominator.

Objective 1 Multiply radical expressions.

Multiply. Assume that all variables represent nonnegative real numbers.

1. $-3\sqrt{6}\cdot 8\sqrt{50}$ **1.**________________

2. $\sqrt{35}\cdot 6\sqrt{15}$ **2.**________________

3. $10\sqrt[3]{18}\cdot 7\sqrt[3]{21}$ **3.**________________

Name: Date:
Instructor: Section:

4. $-2\sqrt[3]{6} \cdot 13\sqrt[3]{44}$ 4.__________

5. $\sqrt{42x^7} \cdot \sqrt{70x^3}$ 5.__________

Objective 2 Use the distributive property to multiply radical expressions.

Multiply. Assume that all variables represent nonnegative real numbers.

6. $\sqrt[3]{20}\left(\sqrt[3]{22} - \sqrt[3]{225}\right)$ 6.__________

7. $\sqrt[3]{9}\left(\sqrt[3]{39} + \sqrt[3]{45}\right)$ 7.__________

8. $-9\sqrt{14}\left(8\sqrt{7} - 3\sqrt{8}\right)$ 8.__________

9. $-5\sqrt{7}\left(2\sqrt{21} + 13\sqrt{28}\right)$ 9.__________

10. $8\sqrt{2}\left(3\sqrt{2} + 5\sqrt{10} - 9\sqrt{14}\right)$ 10.__________

Objective 3 Multiply radical expressions that have two or more terms.

Multiply.

11. $\left(4\sqrt{5} - 8\sqrt{6}\right)\left(8\sqrt{6} + 4\sqrt{5}\right)$ 11.__________

Name: Date:
Instructor: Section:

12. $\left(2\sqrt{22}-3\sqrt{15}\right)\left(8\sqrt{33}-11\sqrt{10}\right)$

12.________________

13. $\left(5+\sqrt{7}\right)\left(3-\sqrt{7}\right)$

13.________________

14. $\left(7\sqrt{10}-6\right)\left(7\sqrt{2}-12\sqrt{5}\right)$

14.________________

15. $\left(\sqrt{6}+15\sqrt{2}\right)^2$

15.________________

16. $\left(\sqrt{6}+\sqrt{5}\right)\left(\sqrt{6}-\sqrt{5}\right)$

16.________________

Objective 4 Multiply radical expressions that are conjugates.

Multiply the conjugates.

17. $\left(\sqrt{6}-\sqrt{10}\right)\left(\sqrt{6}+\sqrt{10}\right)$

17.________________

18. $\left(\sqrt{8}-\sqrt{7}\right)\left(\sqrt{8}+\sqrt{7}\right)$

18.________________

19. $\left(4+2\sqrt{3}\right)\left(4-2\sqrt{3}\right)$

19.________________

Name: Date:
Instructor: Section:

20. $\left(5+4\sqrt{2}\right)\left(5-4\sqrt{2}\right)$ **20.**________________

Objective 5 Rationalize a denominator that has one term.

Simplify. Assume that all variables represent nonnegative real numbers.

21. $\sqrt{\dfrac{28}{7}}$ **21.**________________

22. $\sqrt[3]{\dfrac{540}{10}}$ **22.**________________

23. $\dfrac{\sqrt[4]{240}}{\sqrt[4]{5}}$ **23.**________________

24. $\sqrt[5]{\dfrac{x^{17}y^{6}z^{10}}{x^{4}y^{16}z^{8}}}$ **24.**________________

25. $\sqrt{\dfrac{6}{x^{14}y^{8}}}$ **25.**________________

Rationalize the denominator and simplify. Assume that all variables represent nonnegative real numbers.

26. $\sqrt{\dfrac{7}{2}}$ **26.**________________

Name: Date:
Instructor: Section:

27. $\frac{3}{\sqrt{6}}$

27.________________

28. $\sqrt[3]{\frac{16}{9}}$

28.________________

29. $\sqrt[3]{\frac{27}{28}}$

29.________________

30. $\frac{x^2 y^4}{\sqrt[4]{x^7 y^{14}}}$

30.________________

Objective 6 Rationalize a denominator that has two terms.

Rationalize the denominator.

31. $\frac{2\sqrt{6}}{7\sqrt{2}+3\sqrt{3}}$

31.________________

32. $\frac{4\sqrt{10}}{3\sqrt{11}+2\sqrt{5}}$

32.________________

33. $\frac{\sqrt{14}+5\sqrt{2}}{2\sqrt{7}+\sqrt{2}}$

33.________________

Name: Date:
Instructor: Section:

34. $\dfrac{2\sqrt{6}+10}{8\sqrt{3}-5}$

34.________________

35. $\dfrac{7\sqrt{12}+14\sqrt{11}}{3\sqrt{121}-\sqrt{144}}$

35.________________

36. $\dfrac{22\sqrt{6}-33\sqrt{7}}{5\sqrt{36}+2\sqrt{49}}$

36.________________

37. $\dfrac{15}{\sqrt{22}+\sqrt{12}}$

37.________________

38. $\dfrac{\sqrt{2}}{5\sqrt{14}+4\sqrt{6}}$

38.________________

Name: Date:
Instructor: Section:

Chapter 6 RADICAL EXPRESSIONS AND EQUATIONS

6.5 Radical Equations and Applications of Radical Equations

Learning Objectives

1. Solve radical equations.
2. Solve equations containing radical functions.
3. Solve equations containing rational exponents.
4. Solve equations in which a radical is equal to a variable expression.
5. Solve equations containing two radicals.
6. Solve applied problems involving a pendulum and its period.
7. Solve other applied problems involving radicals.

Objective 1 Solve radical equations.

Solve. (Check for extraneous solutions.)

1. $\sqrt{2x+11}=-4$ **1.**________________

2. $\sqrt{x^2-13x+40}=2$ **2.**________________

3. $\sqrt{x+7}=13$ **3.**________________

4. $\sqrt{4x+1}-1=2$ **4.**________________

5. $\sqrt[4]{x+2}+2=3$ **5.**________________

Objective 2 Solve equations containing radical functions.

Solve. (Check for extraneous solutions.)

6. For the function $f(x)=\sqrt{x^2-x+5}$, find all values x for which $f(x)=5$. **6.**________________

Name: Date:
Instructor: Section:

7. For the function $f(x)=\sqrt{3x+4}$, find all values x for which $f(x)=4$.

7.______________

8. For the function $f(x)=\sqrt{x^2-7x+34}$, find all values x for which $f(x)=8$.

8.______________

9. For the function $f(x)=\sqrt{x^2-4x+49}-2$, find all values x for which $f(x)=7$.

9.______________

10. For the function $f(x)=\sqrt{x^2-9x+36}-3$, find all values x for which $f(x)=1$.

10.______________

Objective 3 Solve equations containing rational exponents.

Solve. (Check for extraneous solutions.)

11. $x^{1/2}-2=7$

11.______________

12. $x^{1/3}+5=-1$

12.______________

Name: Date:
Instructor: Section:

Objective 4 Solve equations in which a radical is equal to a variable expression.

Solve. (Check for extraneous solutions.)

13. $\sqrt{13x-36}=x$ **13.**______________

14. $\sqrt{9x+55}=x+5$ **14.**______________

15. $\sqrt{7x+36}-4=x$ **15.**______________

16. $\sqrt{51-7x}+9=x$ **16.**______________

17. $\sqrt{5x+21}=2x+6$ **17.**______________

Objective 5 Solve equations containing two radicals.

Solve. (Check for extraneous solutions.)

18. $\sqrt{x^2-x-12}=\sqrt{x^2-4x+3}$ **18.**______________

19. $\sqrt{2x+7}-\sqrt{x}=2$ **19.**______________

Name: Date:
Instructor: Section:

20. $\sqrt{x^2+12x-52}=\sqrt{7x-16}$

20.________________

21. $\sqrt{9x+1}=\sqrt{7x+9}$

21.________________

22. $\sqrt{x^2-2x+3}=\sqrt{2x^2-4x-5}$

22.________________

23. $3+\sqrt{x-2}=\sqrt{x+13}$

23.________________

24. $\sqrt{x+4}+\sqrt{3x+16}=2$

24.________________

Objective 6 Solve applied problems involving a pendulum and its period.

For exercises 1-4 use the formula $T=2\pi\sqrt{\frac{L}{32}}$.

25. If a pendulum has a period of 4 seconds, find its length in feet. Round to the nearest tenth of a foot.

25.________________

Name: Date:
Instructor: Section:

26. If a pendulum has a period of 5 seconds, find its length in feet. Round to the nearest tenth of a foot.

26.________________

27. If the length of a pendulum is 10 feet, what is its period? Round to the nearest hundredth of a foot.

27.________________

28. What is the length of a pendulum that has a period of 3 seconds? Round to the nearest tenth of a foot.

28.________________

If a pendulum has a length of L inches, then its period in seconds, T, can be found using the formula $T = 2\pi\sqrt{\frac{L}{384}}$.

29. A pendulum has a length of 109 inches. Find its period, rounded to the nearest hundredth of a second.

29.________________

30. If a pendulum has a period of 1.2 seconds, find its length in inches. Round to the nearest tenth of an inch.

30.________________

Name: Date:
Instructor: Section:

If a pendulum has a length of L meters, then its period in seconds, T, can be found using the formula $T = 2\pi\sqrt{\frac{L}{9.8}}$.

31. If the length of a pendulum is 5 meters, what is its period? Round to the nearest hundredth.

31.________________

Objective 7 Solve other applied problems involving radicals.

Skid mark analysis is one way to estimate the speed a car was traveling prior to an accident. The speed, s, that the vehicle was traveling in miles per hour can be approximated by the formula $s = \sqrt{30df}$, *where d represents the length of the skid marks in feet and f represents the drag factor of the road.*

32. A vehicle involved in an accident made 135 feet of skid marks before crashing. If the drag factor for the road is 0.8, find the speed the car was traveling. Round to the nearest whole number.

32.________________

33. A vehicle that was involved in an accident on a concrete road was traveling at a speed of 55 miles per hour when it started skidding. If the drag factor for concrete is 0.8, find the length of the skid marks made by the car.

33.________________

Name: Date:
Instructor: Section:

A water tank has a hole at the bottom, and the rate r at which water flows out of the hole in gallons per minute can be found using the formula $r = 19.8\sqrt{d}$, *where d represents the depth of the water in the tank in feet.*

34. If water is flowing out of the tank at a rate of 75 gallons per minute, find the depth of water in the tank. Round to the nearest tenth of a foot.

34.________________

35. If water is flowing out of the tank at a rate of 55 gallons per minute, find the depth of water in the tank. Round to the nearest tenth of a foot.

35.________________

36. Find the rate of water flow if the depth of water in the tank is 34 feet. Round to the nearest tenth.

36.________________

37. Find the rate of water flow if the depth of water in the tank is 31 feet. Round to the nearest tenth.

37.________________

A tsunami is a great sea wave caused by natural phenomena such as earthquakes or volcanic activity. The speed of a tsunami, s, in miles per hour at any particular point, can be found from the formula $s = 308.29\sqrt{d}$, *where d is the depth of the ocean in miles at that point.*

38. Find the speed of a tsunami if the depth of the water is 3.4 miles. Round to the nearest whole number.

38.________________

Name: Date:
Instructor: Section:

39. Find the speed of a tsunami if the depth of the water is 4.7 miles. Round to the nearest whole number.

39.________________

40. If a tsunami is traveling at 240 miles per hour, what is the depth of the ocean at that location? Round to the nearest tenth.

40.________________

Name: Date:
Instructor: Section:

Chapter 6 RADICAL EXPRESSIONS AND EQUATIONS

6.6 The Complex Numbers

Learning Objectives

1. Rewrite square roots of negative numbers as imaginary numbers.
2. Add and subtract complex numbers.
3. Multiply imaginary numbers.
4. Multiply complex numbers.
5. Divide by a complex number.
6. Divide by an imaginary number.
7. Simplify expressions containing powers of *i*.

Key Terms

Use the most appropriate term or phrase from the given list to complete each statement in exercises 1-3.

$\boldsymbol{a}$ $\boldsymbol{b}$ $\boldsymbol{i}$ $\sqrt{-1}$ -1

1. For the complex number $a + bi$, the number ________________ is the real part.

2. For the complex number $a + bi$, the number ________________ is the imaginary part.

3. The imaginary unit ________________ is a number that is equal to $\sqrt{-1}$.

Objective 1 Rewrite square roots of negative numbers as imaginary numbers.

Express in terms of i.

1. $\sqrt{-126}$ **1.**________________

2. $-\sqrt{-363}$ **2.**________________

3. $\sqrt{-64}$ **3.**________________

4. $-\sqrt{-294}$ **4.**________________

Name: Date:
Instructor: Section:

Objective 2 Add and subtract complex numbers.

Add or subtract the complex numbers.

5. $(-6+14i)+(8-20i)-(3+11i)$ **5.**____________

6. $(19+7i)-(4-15i)+(-21+16i)$ **6.**____________

7. $(16+13i)+(17-5i)$ **7.**____________

Find the missing complex number.

8. $(8-13i)-?=5-21i$ **8.**____________

9. $?-(17-30i)=-12-17i$ **9.**____________

Objective 3 Multiply imaginary numbers.

Multiply.

10. $-10i\cdot 34i$ **10.**____________

11. $\sqrt{-25}\cdot\sqrt{-81}$ **11.**____________

12. $\sqrt{-1}\cdot\sqrt{-100}$ **12.**____________

Name: Date:
Instructor: Section:

13. $\sqrt{-81} \cdot \sqrt{-36}$ **13.**______________

14. $-5i \cdot 6i$ **14.**______________

Find two imaginary numbers with the given product. (Answers may vary.)

15. -32 **15.**______________

16. -36 **16.**______________

Objective 4 Multiply complex numbers.

Multiply.

17. $-12i(16+5i)$ **17.**______________

18. $(8-7i)(1+6i)$ **18.**______________

19. $(5+12i)^2$ **19.**______________

20. $(11-8i)(11+8i)$ **20.**______________

Name: Date:
Instructor: Section:

21. $(5-12i)^2$ **21.**______________

Multiply the conjugates, using the fact that $(a+bi)(a-bi)=a^2+b^2$.

22. $(-3-4i)(-3+4i)$ **22.**______________

23. $(-2+9i)(-2-9i)$ **23.**______________

Objective 5 Divide by a complex number.

Rationalize the denominator.

24. $\frac{4-10i}{2-9i}$ **24.**______________

25. $\frac{6+5i}{8+3i}$ **25.**______________

Objective 6 Divide by an imaginary number.

Rationalize the denominator.

26. $\frac{5-12i}{4i}$ **26.**______________

Name: Date:
Instructor: Section:

27. $\dfrac{10+15i}{7i}$

27. ________________

Divide.

28. $(7-5i)\div(-2i)$

28. ________________

29. $5\div(3+i)$

29. ________________

Find the missing complex number.

30. $\dfrac{12+5i}{?}=2+3i$

30. ________________

31. $\dfrac{68-17i}{?}=4+i$

31. ________________

Objective 7 Simplify expressions containing powers of *i*.

32. i^{11}

32. ________________

33. i^{20}

33. ________________

Name: Date:
Instructor: Section:

34. i^{33}

34._______________

35. $i^3 + i^5$

35._______________

Name: Date:
Instructor: Section:

Chapter 7 QUADRATIC EQUATIONS

7.1 Solving Quadratic Equations by Extracting Square Roots; Completing the Square

Learning Objectives

1. Solve quadratic equations by factoring.
2. Solve quadratic equations by extracting square roots.
3. Solve quadratic equations by extracting square roots involving a linear expression that is squared.
4. Solve applied problems by extracting square roots.
5. Solve quadratic equations by completing the square.
6. Find a quadratic equation, given its solutions.

Key Terms

Use the most appropriate term or phrase from the given list to complete each statement in exercises 1-2.

linear **squared** **extracting square roots** **completing the square**

1. To solve a quadratic equation by extracting square roots, begin by isolating the ______________________ term.

2. The method of ______________________ can be used to solve quadratic equations having both a first-degree term and second-degree term.

Objective 1 Solve quadratic equations by factoring.

Solve by factoring.

1. $x^2 - 4x - 12 = 0$ **1.**________________

2. $x^2 - 11x + 24 = 0$ **2.**________________

3. $x^2 - 36 = 0$ **3.**________________

Name: Date:
Instructor: Section:

4. $3x^2 - 7x - 6 = 0$ **4.**________________

5. $x^2 + 20x + 76 = 4x + 13$ **5.**________________

6. $x^2 + 8x - 20 = 10(x + 6)$ **6.**________________

Objective 2 Solve quadratic equations by extracting square roots.

Solve by extracting square roots.

7. $x^2 = 25$ **7.**________________

8. $x^2 = 100$ **8.**________________

9. $x^2 - 20 = 0$ **9.**________________

10. $x^2 + 12 = 0$ **10.**________________

Name: Date:
Instructor: Section:

Objective 3 Solve quadratic equations by extracting square roots involving a linear expression that is squared.

Solve by extracting square roots.

11. $(2x-3)^2=24$ **11.**________________

12. $(3x+2)^2+11=23$ **12.**________________

13. $(x+4)^2=81$ **13.**________________

14. $(x-3)^2+7=23$ **14.**________________

15. $(2x+1)^2-17=19$ **15.**________________

Objective 4 Solve applied problems by extracting square roots.

Approximate to the nearest tenth when necessary.

16. The area of a square is 49 square yards. Find the length of a side of the square. **16.**________________

Name: Date:
Instructor: Section:

17. The area of a square is 72 square meters. Find the length of a side of the square.

17.________________

18. The area of a circle is 28 square centimeters. Find the radius of the circle.

18.________________

19. The area of a circle is 112 square inches. Find the radius of the circle.

19.________________

Objective 5 Solve quadratic equations by completing the square.

Fill in the missing term that makes the expression a perfect square trinomial, and factor the resulting expression.

20. $x^2 + 7x +$ ___

20.________________

21. $x^2 + 20x +$ ___

21.________________

22. $x^2 + \frac{4}{9}x +$ ___

22.________________

23. $x^2 - \frac{7}{2}x +$ ___

23.________________

Name: Date:
Instructor: Section:

Solve by completing the square.

24. $3x^2 + 8x + 4 = 0$ **24.**________________

25. $x^2 - 12x + 35 = 0$ **25.**________________

26. $x^2 - 2x - 15 = 0$ **26.**________________

27. $x^2 + \frac{4}{5}x - 2 = 0$ **27.**________________

28. $x^2 - \frac{1}{3}x - \frac{10}{3} = 0$ **28.**________________

29. $x^2 + 14x + 48 = 0$ **29.**________________

30. $2x^2 - 8x - 15 = 0$ **30.**________________

Objective 6 Find a quadratic equation, given its solutions.

Find a quadratic equation with integer coefficients that has the following solution set.

31. $\{2+\sqrt{5},\ 2-\sqrt{5}\}$ **31.**________________

Name: Date:
Instructor: Section:

32. $\left\{-3+2\sqrt{3},\ -3-2\sqrt{3}\right\}$ **32.**________________

33. $\{0,\ -2\}$ **33.**________________

34. $\{6,\ -6\}$ **34.**________________

35. $\left\{3,\ -\frac{1}{2}\right\}$ **35.**________________

36. $\left\{7\sqrt{3},\ -7\sqrt{3}\right\}$ **36.**________________

37. $\left\{4+\sqrt{3},\ 4-\sqrt{3}\right\}$ **37.**________________

38. $\{2+i,\ 2-i\}$ **38.**________________

Name: Date:
Instructor: Section:

Chapter 7 QUADRATIC EQUATIONS

7.2 The Quadratic Formula

Learning Objectives
1 Derive the quadratic formula.
2 Identify the coefficients a, b, and c of a quadratic equation.
3 Solve quadratic equations by the quadratic formula.
4 Use the discriminant to determine the number and type of solutions of a quadratic equation.
5 Use the discriminant to determine whether a quadratic expression is factorable.
6 Solve projectile motion problems.

Key Terms
Use the most appropriate term or phrase from the given list to complete each statement in exercises 1-3.

zero	**one**	**two**	**positive**	**negative**

1. A quadratic equation has one real solution if the value of the discriminant is ____________________.

2. A quadratic equation has no real solutions if the value of the discriminant is ____________________.

3. A quadratic equation has two real solutions if the value of the discriminant is ____________________.

Objective 1 Derive the quadratic formula.
Objective 2 Identify the coefficients a, b, and c of a quadratic equation.
Objective 3 Solve quadratic equations by the quadratic formula.

Solve by using the quadratic formula.

1. $x^2+13x+31=0$ **1.**________________

2. $x^2-8x+11=0$ **2.**________________

Name: Date:
Instructor: Section:

3. $x^2 - 17 = 8x$ **3.**________________

4. $x^2 + 24 = 10x$ **4.**________________

5. $16x^2 + 8x + 1 = 0$ **5.**________________

6. $9x^2 - 12x + 4 = 0$ **6.**________________

7. $x^2 - 8x + 12 = 0$ **7.**________________

8. $x^2 - 4x - 45 = 0$ **8.**________________

9. $x^2 + 9x + 25 = 0$ **9.**________________

Name: Date:
Instructor: Section:

10. $4x^2 - 16x - 9 = 0$ **10.**________________

11. $(x+6)(x+8) = x + 26$ **11.**________________

12. $12x^2 + 4x + 5 = 3x^2 + 22x + 3$ **12.**________________

Objective 4 Use the discriminant to determine the number and type of solutions of a quadratic equation.

For the following quadratic equations, use the discriminant to determine the number and type of solutions.

13. $x^2 - 10x + 39 = 0$ **13.**________________

14. $24x^2 - 49x + 15 = 0$ **14.**________________

15. $x^2 - x = 42$ **15.**________________

16. $x^2 + 7x = -20$ **16.**________________

Name: Date:
Instructor: Section:

17. $x^2 + 3x + 15 = 0$ **17.**________________

18. $x^2 - 18x + 81 = 0$ **18.**________________

19. $2x^2 + 16x - 11 = 0$ **19.**________________

20. $3x^2 + 14x + 21 = 0$ **20.**________________

21. $x^2 + 8x + 16 = 0$ **21.**________________

22. $x^2 - 5x - 52 = 0$ **22.**________________

Objective 5 Use the discriminant to determine whether a quadratic expression is factorable.

Use the discriminant to determine whether each of the following quadratic expressions is factorable or prime.

23. $x^2 + 10x + 16$ **23.**________________

24. $x^2 - 4x + 9$ **24.**________________

Name: Date:
Instructor: Section:

25. $x^2 + 3x - 6$ **25.**________________

26. $8x^2 - 10x - 3$ **26.**________________

27. $15x^2 - 29x + 12$ **27.**________________

28. $13x^2 + 12x - 6$ **28.**________________

29. $8x^2 + 25x - 10$ **29.**________________

30. $16x^2 - 62x - 45$ **30.**________________

Solve the following quadratic equations using the most efficient technique for each (factoring, extracting square roots, completing the square, or quadratic formula).

31. $x^2 - 18x + 57 = 0$ **31.**________________

32. $x^2 = 84$ **32.**________________

33. $x^2 - 4x - 32 = 0$ **33.**________________

Name: Date:
Instructor: Section:

34. $x^2 - 7x + 30 = 0$ **34.**________________

35. $6x^2 - 19x + 15 = 0$ **35.**________________

36. $x^2 - 11x + 24 = 0$ **36.**________________

Objective 6 Solve projectile motion problems.

Solve using the function $h(t) = -16t^2 + v_0 t + s$. *Approximate to the nearest tenth when necessary.*

37. A projectile is launched from the roof of a building 117 feet tall, with an initial speed of 120 feet per second. How long will it take until the projectile lands on the ground? Round to the nearest tenth of a second. **37.**________________

38. An object is launched upward from the top of a building 36 feet high, with an initial velocity of 139 feet per second. At what time(s) is the object 263 feet above the ground? Round to the nearest tenth of a second. **38.**________________

Name: Date:
Instructor: Section:

Chapter 7 QUADRATIC EQUATIONS

7.3 Equations That Are Quadratic in Form

Learning Objectives
1 Solve equations by making a u-substitution.
2 Solve radical equations.
3 Solve rational equations.
4 Solve work-rate problems.

Objective 1 Solve equations by making a u-substitution.

Solve by making a u-substitution.

1. $x^4 - 8x^2 = 9$ **1.**________________

2. $x^4 - 3x^2 = 4$ **2.**________________

3. $x^4 - 10x^2 + 24 = 4x^2$ **3.**________________

4. $x^4 + 9x^2 - 16 = 13x^2 + 16$ **4.**________________

5. $x - 3\sqrt{x} - 18 = 0$ **5.**________________

Name: Date:
Instructor: Section:

6. $x+7\sqrt{x}=-12$ **6.**________________

7. $x-x^{1/2}=42$ **7.**________________

8. $x+7x^{1/2}=-10$ **8.**________________

9. $x^{2/3}-9x^{1/3}-22=0$ **9.**________________

10. $x^{2/3}-10x^{1/3}+21=0$ **10.**________________

11. $\left(x^2+2x\right)^2-7\left(x^2+2x\right)-8=0$ **11.**________________

12. $(2x+8)^2+9(2x+8)+14=0$ **12.**________________

Name: Date:
Instructor: Section:

Objective 2 Solve radical equations.

Solve.

13. $\sqrt{2x+1}=7$

13.________________

14. $\sqrt{3x+7}=5$

14.________________

15. $\sqrt{x^2+5x-8}=\sqrt{x+13}$

15.________________

16. $\sqrt{x^2-4x-5}=\sqrt{6x-21}$

16.________________

17. $\sqrt{3x^2+x-20}=\sqrt{2x^2+7x+7}$

17.________________

18. $\sqrt{2x+9}+3=x$

18.________________

19. $\sqrt{x+4}+1=\sqrt{3x+1}$

19.________________

Name: Date:
Instructor: Section:

Objective 3 Solve rational equations.

Solve.

20. $x+\frac{35}{x}=12$

20.________________

21. $x=\frac{48}{x-2}$

21.________________

22. $1+\frac{5}{x}-\frac{24}{x^2}=0$

22.________________

23. $\frac{3}{x}-\frac{5}{x^2}=-2$

23.________________

24. $1-\frac{6}{x}-\frac{7}{x^2}=0$

24.________________

25. $x=\frac{60}{x+4}$

25.________________

Name: Date:
Instructor: Section:

26. $\dfrac{x+6}{x+4}+\dfrac{5}{x-5}=\dfrac{6}{x^2-x-20}$

26.________________

27. $\dfrac{x}{x+1}+\dfrac{1}{x-7}=\dfrac{17}{x^2-6x-7}$

27.________________

28. $x+5=\dfrac{14}{x}$

28.________________

Objective 4 Solve work-rate problems.

Solve.

29. Two drainpipes, working together, can drain a swimming pool in 25 hours. The larger pipe, working alone, can drain the pool in 10 hours less than the smaller pipe. How long would it take the larger pipe to drain the tank on its own? Round to the nearest tenth of an hour.

29.________________

Name: Date:
Instructor: Section:

30. Working together, two people can cut a large lawn in 4 hours. One person can do the job alone in 1 hour less than the other person. How long would it take the faster person to do the job alone? Round to the nearest tenth of an hour.

30._______________

Name: Date:
Instructor: Section:

Chapter 7 QUADRATIC EQUATIONS

7.4 Graphing Quadratic Equations and Quadratic Functions

Learning Objectives
1 Graph quadratic equations.
2 Graph parabolas that open downward.
3 Graph quadratic functions of the form $f(x) = ax^2 + bx + c$.
4 Find the vertex of a parabola by completing the square.
5 Graph quadratic equations of the form $y = a(x - h)^2 + k$.
6 Graph quadratic equations of the form $f(x) = a(x - h)^2 + k$

Key Terms
Use the most appropriate term from the given list to complete each statement in exercises 1-3.

vertex **axis of symmetry** **symmetric** ***y*-intercepts** ***x*-intercepts**

1. A parabola is always ________________.

2. The ________________ goes through the vertex.

3. If a parabola opens upward and has its vertex above the x-axis, then there are no ________________.

Objective 1 Graph quadratic equations.

Find the vertex of the parabola associated with each of the following quadratic equations, as well as the equation of the axis of symmetry.

1. $y = 4x^2 - 16x + 27$ **1.**________________

2. $y = -4x^2 + 16x - 19$ **2.**________________

Name: Date:
Instructor: Section:

3. $y=-4x^2-7x+4$ **3.**________________

4. $y=(x+1)^2+6$ **4.**________________

5. $y=x^2-9x-11$ **5.**________________

Find the x-and y-intercepts of the parabola associated with the given quadratic equations. If necessary, round to the nearest tenth. If the parabola does not have any x-intercepts, state "no x-intercepts."

6. $y=-x^2+3x+18$ **6.**________________

7. $y=x^2+6x-16$ **7.**________________

8. $y=-x^2-11x-30$ **8.**________________

Name: Date:
Instructor: Section:

9. $y = -x^2 + 6x - 4$

9.________________

10. $y = -x^2 + 5x + 66$

10.________________

11. $y = (x - 7)^2 + 3$

11.________________

Graph the given parabolas. Find the vertex and all intercepts.

12. $y = x^2 - 7x + 5$

12.________________

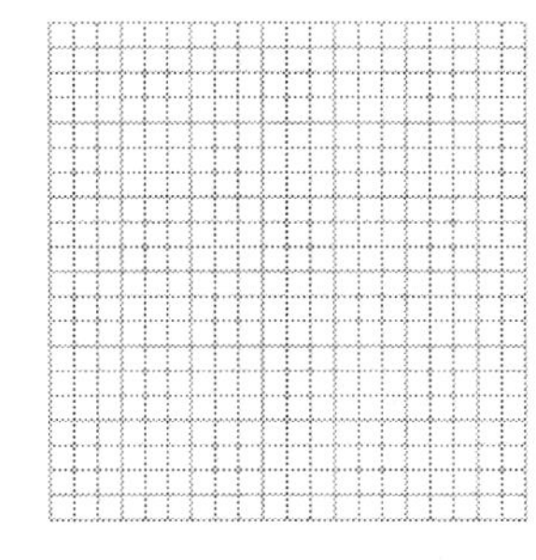

13. $y = x^2 - 5x - 24$

13.________________

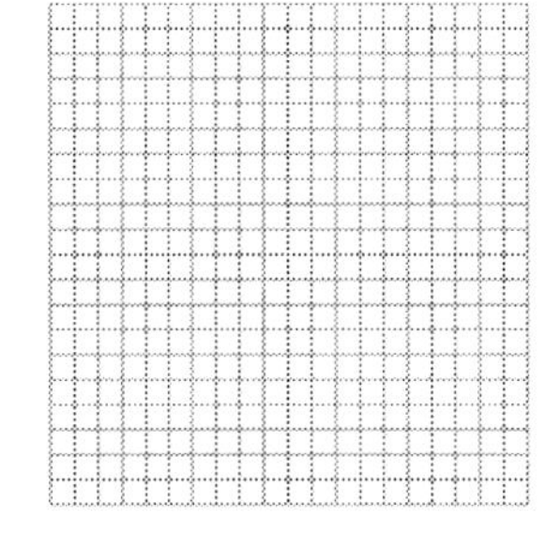

Name: Date:
Instructor: Section:

14. $y = x^2 - 7x + 10$

14.________________

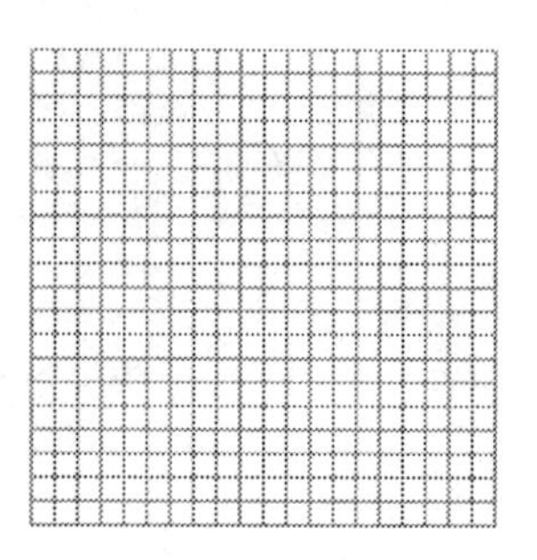

Objective 2 Graph parabolas that open downward.

Graph the following parabolas. Find the vertex and all intercepts.

15. $y = -x^2 + 3x - 4$

15.________________

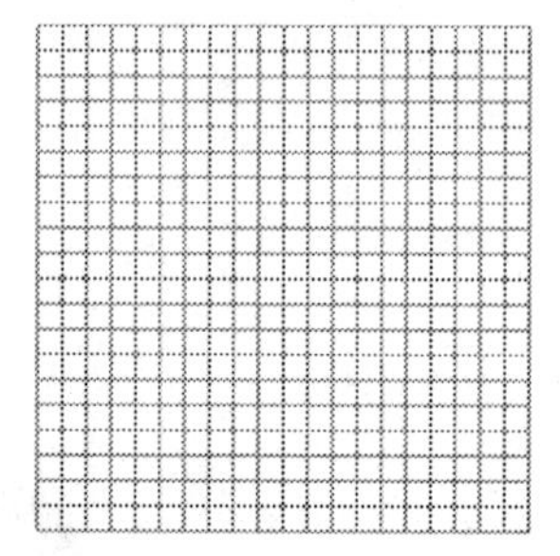

16. $y = -x^2 + 4$

16.________________

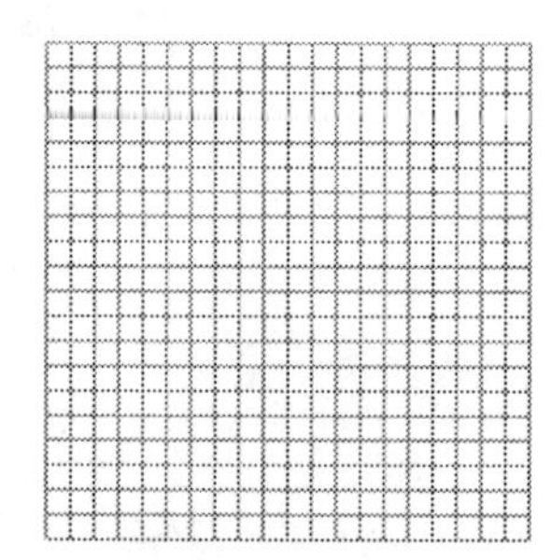

17. $y = -x^2 + 5x + 6$

17.________________

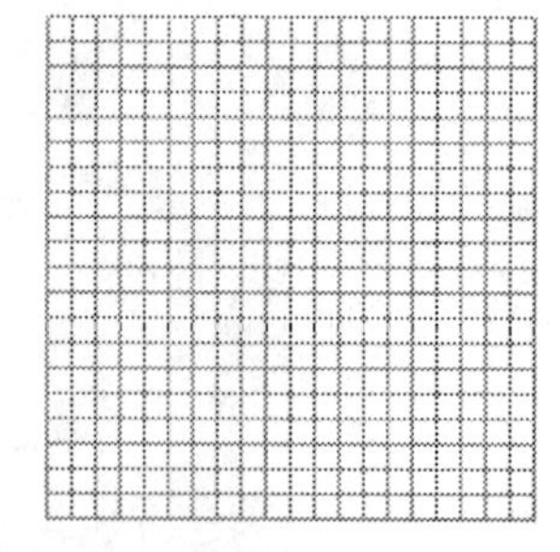

Name: Date:
Instructor: Section:

Objective 3 Graph quadratic functions of the form $f(x) = ax^2 + bx + c$.

Graph the given parabolas. Find the vertex and all intercepts.

18. $f(x) = x^2 + 4x - 1$

18.________________

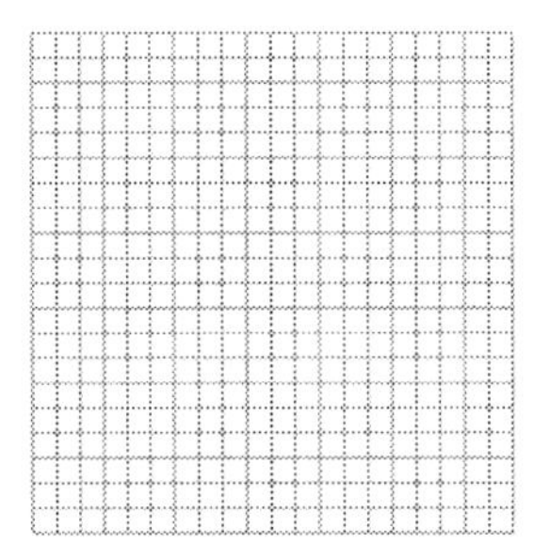

19. $f(x) = x^2 - 2x + 6$

19.________________

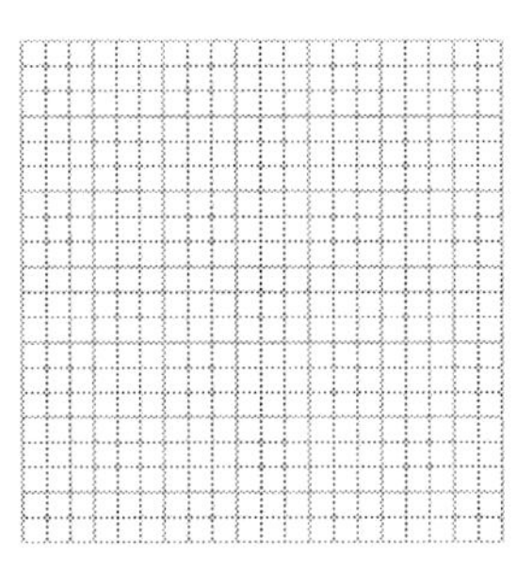

20. $f(x) = -x^2 + 7x - 19$

20.________________

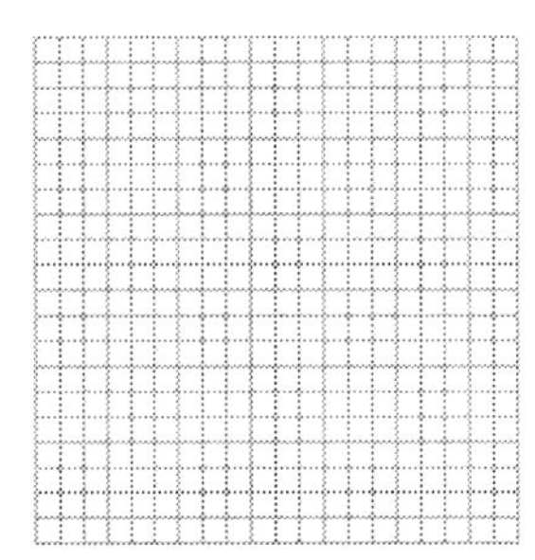

21. $f(x) = -x^2 - 5x + 3$

21.________________

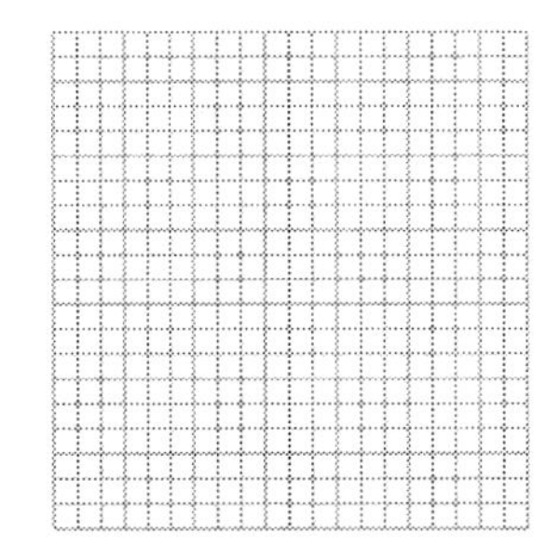

Name: Date:
Instructor: Section:

Objective 4 Find the vertex of a parabola by completing the square.

Find the vertex of each parabola by completing the square.

22. $y = x^2 - 4x + 3$ **22.**________________

23. $y = x^2 - 6x + 5$ **23.**________________

24. $y = x^2 - 6x + 8$ **24.**________________

25. $y = -x^2 + 12x - 34$ **25.**________________

26. $y = -x^2 + 10x - 26$ **26.**________________

Name: Date:
Instructor: Section:

Objective 5 Graph quadratic equations of the form $y = a(x - h)^2 + k$.

Graph the following parabolas. Find the vertex and all intercepts.

27. $y = (x-1)^2 - 4$

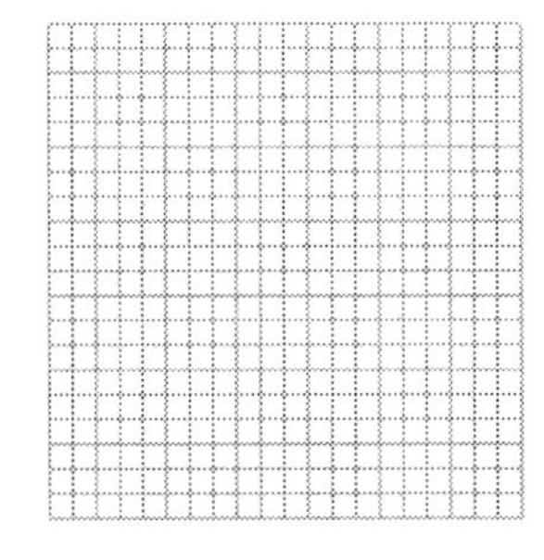

28. $y = -(x-2)^2 + 4$

28.________________

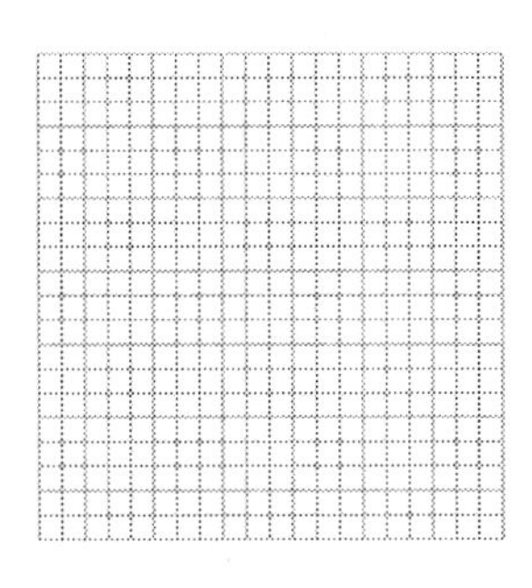

Objective 6 Graph quadratic equations of the form $f(x) = a(x - h)^2 + k$.

Graph the following parabolas. Find the vertex and all intercepts.

29. $f(x) = (x-4)^2 - 36$

29.________________

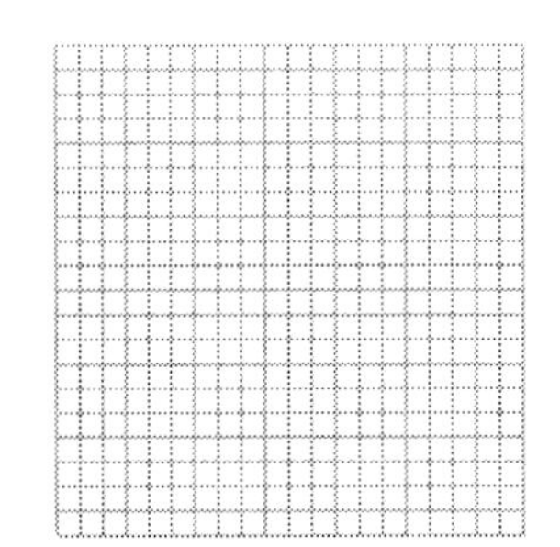

Name: Date:
Instructor: Section:

30. $f(x) = -(x-4)^2 + 5$

30.________________

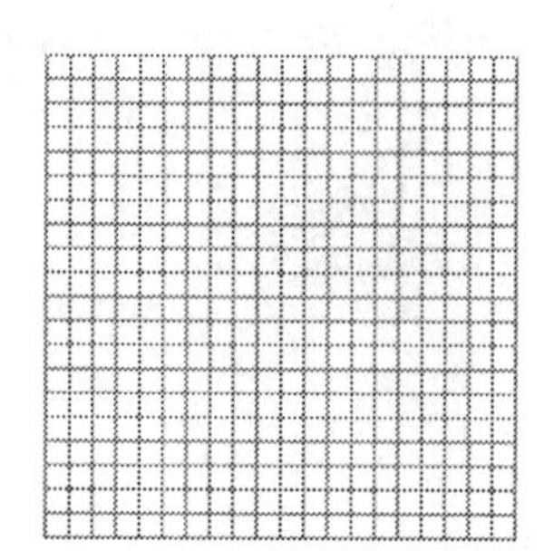

Name: Date:
Instructor: Section:

Chapter 7 QUADRATIC EQUATIONS

7.5 Applications Using Quadratic Equations

Learning Objectives
1 Solve applied geometric problems.
2 Solve problems by using the Pythagorean theorem.
3 Solve applied problems by using the Pythagorean theorem.
4 Find the maximum or minimum value of a quadratic function.
5 Solve applied maximum/minimum problems.

Objective 1 Solve applied geometric problems.

Solve. Approximate to the nearest tenth when necessary.

1. The height of a triangle is 2 feet less than three times its base. If the area of the triangle is 32.5 square feet, find the base and height of the triangle.

1.________________

2. The base of a triangle is three times its height. If the area of the triangle is 54 square inches, find the base and height of the triangle.

2.________________

3. The gable of a house is the triangular wall section at the end of a pitched roof. The base of a house's gable is 3 feet longer than 5 times its height. If the area of the gable is 70 square feet, find the base and height of the gable.

3.________________

Name: Date:
Instructor: Section:

4. The sail on a sailboat is shaped like a triangle, with an area of 31.5 square feet. If the height of the sail is 5 feet longer than twice the base of the sail, find the base and height of the sail.

4.________________

5. A triangular kite is made of 42 square inches of material. If the base of the kite is 2 inches more than twice the height of the kite, find the base and height of the kite.

5.________________

6. The perimeter of a rectangle is 52 feet and its area is 165 square feet. Find the dimensions of this rectangle.

6.________________

7. The perimeter of a rectangle is 20 inches, and the area is 21 square inches. Find the dimensions of the rectangle.

7.________________

Name: Date:
Instructor: Section:

8. A rectangular swimming pool has a perimeter of 104 feet. If the swimming pool covers an area of 640 square feet, find the dimensions of the swimming pool.

8.________________

9. A rectangular horse corral is surrounded by a fence 160 feet long. If the corral covers an area of 1600 square feet, find the dimensions of the corral.

9.________________

10. The width of a rectangular lawn is 24 feet more than its length. If the area of the lawn is 7000 square feet, find the dimensions of the lawn. Round to the nearest tenth of a foot.

10.________________

Objective 2 Solve problems by using the Pythagorean theorem.

Solve. Approximate to the nearest tenth when necessary.

11. A right triangle has a leg that measures 6 feet and a hypotenuse that measures 24 feet. Find the length of the other leg, to the nearest tenth of a foot.

11.________________

Name: Date:
Instructor: Section:

12. A right triangle has a leg that measures 3 meters and a hypotenuse that measures 8 meters. Find the length of the other leg, to the nearest tenth of a meter.

12.________________

13. A right triangle with has one leg that measures 14 feet and another leg that measures 13 feet. Find the length of the hypotenuse to the nearest tenth of a foot.

13.________________

Objective 3 Solve applied problems by using the Pythagorean theorem.

Solve. Approximate to the nearest tenth when necessary.

14. A ladder 20 feet long is leaning on the top of a wall. If the distance from the base of the wall to the bottom of the ladder is 10 feet less than the height of the wall, find the height of the wall. Round to the nearest tenth of a foot.

14.________________

15. The diagonal of a rectangular mural is twice its height. If the width of the mural is 5 feet, how tall is the mural? Round to the nearest tenth of a foot.

15.________________

Name: Date:
Instructor: Section:

16. A room is in the shape of a rectangle. The diagonal of the room is 1 foot more than the length of the room. If the width of the room is 7 feet, find the length of the room.

16.________________

17. The diagonal of a rectangular blanket is 3 feet. If the width of the blanket is 1 foot less than the length of the blanket, find the dimensions of the blanket. Round to the nearest tenth of a foot.

17.________________

18. Patti drove 25 miles to the west, and then she drove 60 miles to the south. How far is she from her starting location?

18.________________

19. A vertical post is casting a shadow on the ground. If the post is 4 feet taller than the shadow on the ground, and the tip of the shadow is 20 feet from the top of the post, how tall is the post?

19.________________

Name: Date:
Instructor: Section:

20. A guy wire 32 feet long runs from the top of a pole to a spot on the ground. If the height of the pole is 6 feet more than the distance from the base of the pole to the spot where the guy wire is anchored, how tall is the pole? Round to the nearest tenth of a foot.

20._______________

Objective 4 Find the maximum or minimum value of a quadratic function.

Determine whether the given quadratic function has a maximum value or a minimum value. Then find that maximum or minimum value.

21. $f(x) = x^2 + 2x + 325$

21._______________

22. $f(x) = x^2 - 7x + 13$

22._______________

23. $f(x) = -x^2 - 11x + 16$

23._______________

24. $f(x) = -x^2 + 5x - 38$

24._______________

Name: Date:
Instructor: Section:

25. $f(x)=\frac{1}{2}x^2-\frac{3}{5}x+\frac{17}{6}$

25.________________

26. $f(x)=\frac{3}{4}x^2+\frac{5}{9}x+\frac{37}{18}$

26.________________

27. $f(x)=2x^2+24x+365$

27.________________

28. $f(x)=(x-15)^2+108$

28.________________

Objective 5 Solve applied maximum/minimum problems.

Solve.

29. Ellen is the yearbook advisor at a high school. She must set the price for this year's yearbook. She determines that the revenue generated can be approximated by the function $R(x)=-6x^2+240x$, where x represents the price of the yearbook in dollars. What price will generate the maximum revenue?

29.________________

Name: Date:
Instructor: Section:

30. A college theater department is going to put on a musical. The number of people that will attend depends on the admission charge. If the revenue generated by the performance can be approximated by the function $R(x) = -30x^2 + 550x$, where x represents the admission charge in dollars, what charge will generate the maximum revenue?

30.________________

31. A college is hosting an auction to raise money for their scholarship fund, and the planners are trying to decide the cost of admission. The planners have determined that the revenue generated by admission can be approximated by the function $R(x) = -5x^2 + 450x$, where x represents the price of admission in dollars. What admission price will generate the maximum revenue?

31.________________

32. A graphic artist has decided to make and sell t shirts to fans attending a concert. The average cost per shirt in dollars is given by the function $f(x) = 0.00015x^2 - 0.06x + 10.125$, where x is the number of shirts made. For what number of shirts is the average cost per shirt minimized? What is the minimum cost per shirt that is possible?

32.________________

33. Jackie has 400 feet of fencing to make a rectangular garden. What dimensions will make a garden with the maximum area? What is the maximum area possible?

33.________________

Name: Date:
Instructor: Section:

Chapter 7 QUADRATIC EQUATIONS

7.6 Quadratic and Rational Inequalities

Learning Objectives
1. Solve quadratic inequalities.
2. Solve quadratic inequalities graphically.
3. Solve rational inequalities.
4. Solve inequalities involving functions.
5. Solve applied problems involving inequalities.

Objective 1 Solve quadratic inequalities.

Solve each quadratic inequality. Express your solution on a number line, using interval notation.

1. $x^2 + 2x - 8 \geq 0$ **1.**________________

2. $x^2 + 6x - 7 > 0$ **2.**________________

3. $x^2 - 8x + 11 \leq 0$ **3.**________________

4. $x^2 + 14x + 41 \geq 0$ **4.**________________

5. $\frac{1}{6}x^2 + \frac{5}{3}x + 4 \leq 0$ **5.**________________

Name: Date:
Instructor: Section:

6. $-x^2+7x+70>0$

6.________________

7. $x(x+4)<x-3$

7.________________

8. $\frac{1}{10}x^2+\frac{1}{2}x+\frac{1}{5}\geq 0$

8.________________

9. $-x^2-4x+96\leq 0$

9.________________

10. $x^2-3x+15<-14x-9$

10.________________

11. $x^2-7x-18\geq 0$

11.________________

12. $x^2-5x+6<0$

12.________________

13. $x^2 - 8x + 2 \leq 0$

13.________________

14. $x^2 + 12x + 39 < 0$

14.________________

Objective 2 Solve quadratic inequalities graphically.

Solve.

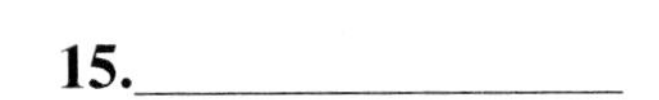

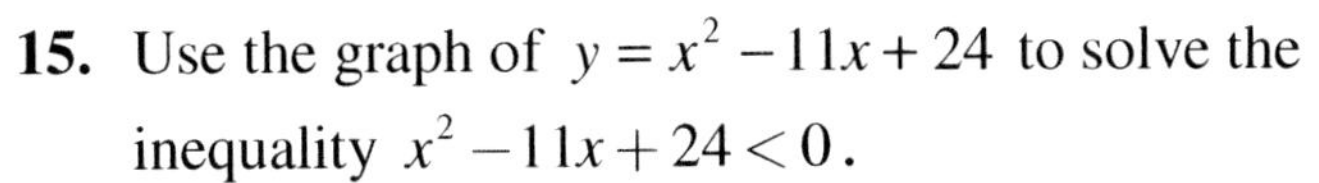

15. Use the graph of $y = x^2 - 11x + 24$ to solve the inequality $x^2 - 11x + 24 < 0$.

15.________________

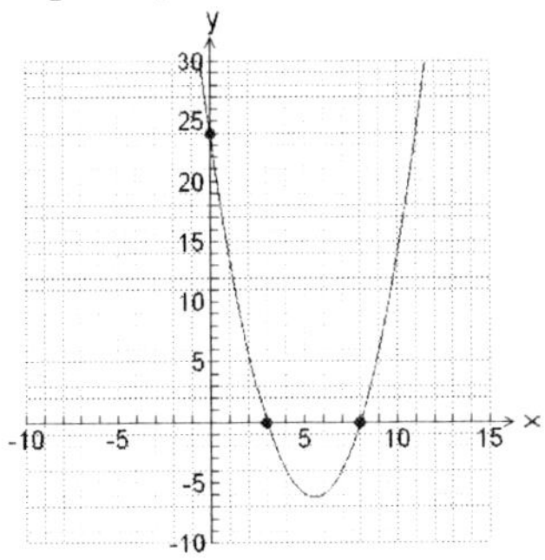

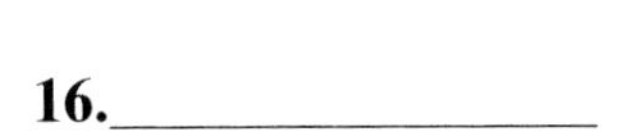

16. Use the graph of $y = -x^2 + 5x + 6$ to solve the inequality $y = -x^2 + 5x + 6 \leq 0$.

16.________________

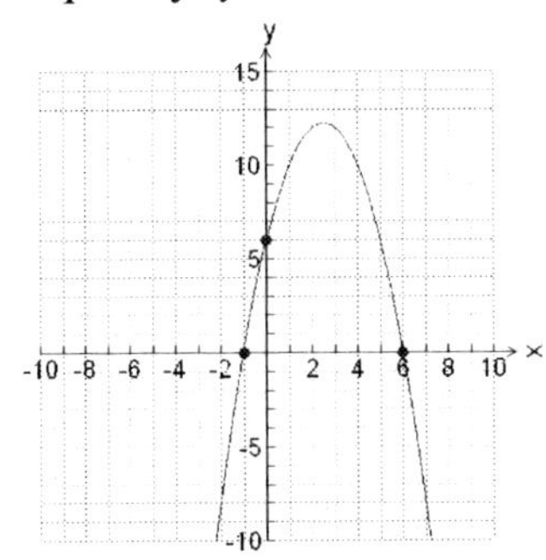

Name: Date:
Instructor: Section:

Objective 3 Solve rational inequalities.

Solve each rational inequality. Express your solution on a number line, using interval notation.

17. $\frac{x(x+2)}{(x-3)(x-3)} > 0$

17.________________

18. $\frac{(x+1)(x-3)(x+5)}{(x+2)(x-4)} \geq 0$

18.________________

19. $\frac{x^2+10x}{x^2+x-30} > 0$

19.________________

20. $\frac{x^2-5x-14}{x^2-4} < 0$

20.________________

21. $\frac{x^2-2x-48}{x^2+7x+12} \geq 0$

21.________________

Name: Date:
Instructor: Section:

22. $\dfrac{x^2-25}{x^3+6x^2+8x}>0$

22.________________

23. $\dfrac{x+9}{x-5}<0$

23.________________

24. $\dfrac{x^2-8x-33}{x^2+6x+8}>0$

24.________________

Objective 4 Solve inequalities involving functions.

Solve.

25. Given $f(x)=x^2+5x+16$, find all values x for which $f(x)>0$.

25.________________

26. Given $f(x)=x^2+15x+56$, find all values x for which $f(x)>0$.

26.________________

Name: Date:
Instructor: Section:

Given the function $f(x)$, solve the inequality $f(x) \geq 0$. Express your solution on a number line using interval notation.

27. $f(x) = -x^2 - 2x + 48$

27.________________

28. $f(x) = \dfrac{x^2 - 3x - 18}{x^2 - 16x + 63}$

28.________________

Objective 5 Solve applied problems involving inequalities.

Solve.

29. A projectile is fired upwards from ground level. After t seconds, its height above the ground is s feet, where $s = -16t^2 + 112t$. For what time period is the projectile at least 160 feet above the ground?

29.________________

30. A projectile is fired upwards from ground level. After t seconds, its height above the ground is s feet, where $s = -16t^2 + 128t$. For what time period is the projectile at least 192 feet above the ground?

30.________________

Name: Date:
Instructor: Section:

Chapter 7 QUADRATIC EQUATIONS

7.7 Other Functions and Their Graphs

Learning Objectives
1. Graph square-root functions.
2. Graph cubic functions.
3. Determine a function from its graph.

Objective 1 Graph square-root functions.

Graph each given square root function, and state its domain and range.

1. $f(x)=\sqrt{2x-1}+5$

1. ________________

____

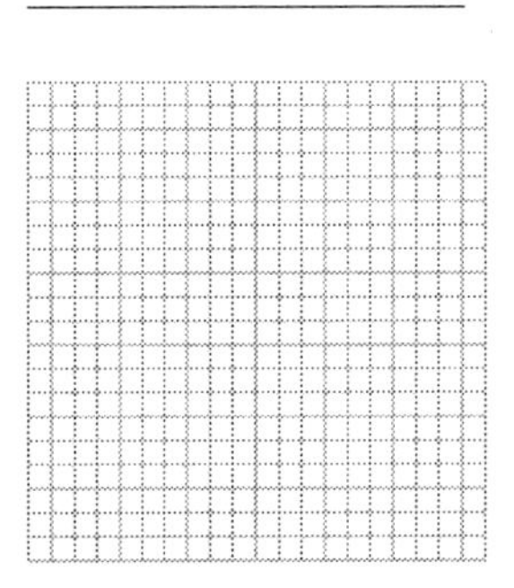

2. $f(x)=\sqrt{4x+3}-1$

2. ________________

____

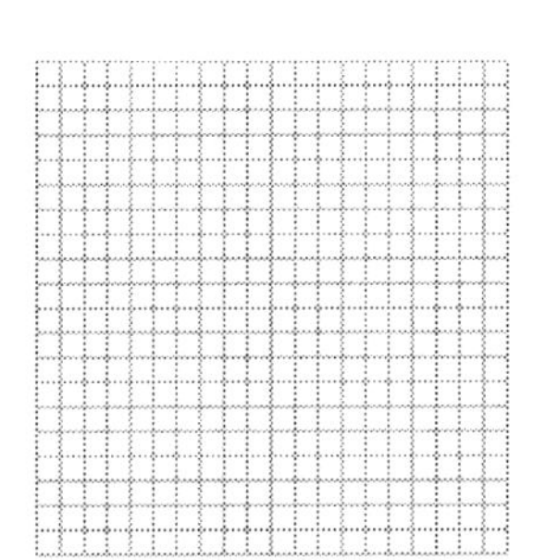

Name: Date:
Instructor: Section:

3. $f(x) = \sqrt{-x}$

3. ______________

____

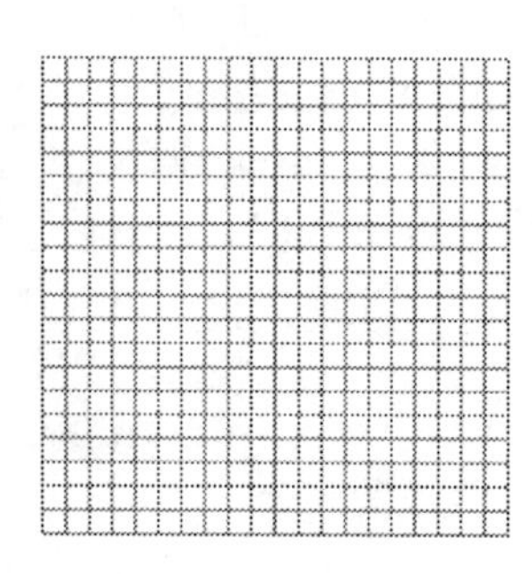

4. $f(x) = \sqrt{6 - x}$

4. ______________

____

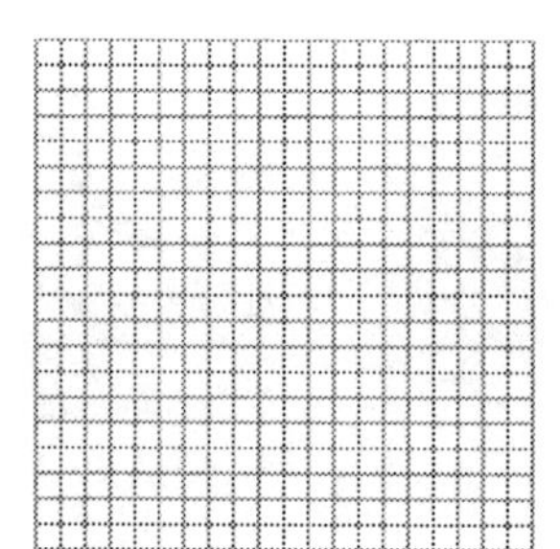

5. $f(x) = \sqrt{x} + 4$

5. ______________

____

Name: Date:
Instructor: Section:

6. $f(x)=\sqrt{9-x}+5$

6.________________

____

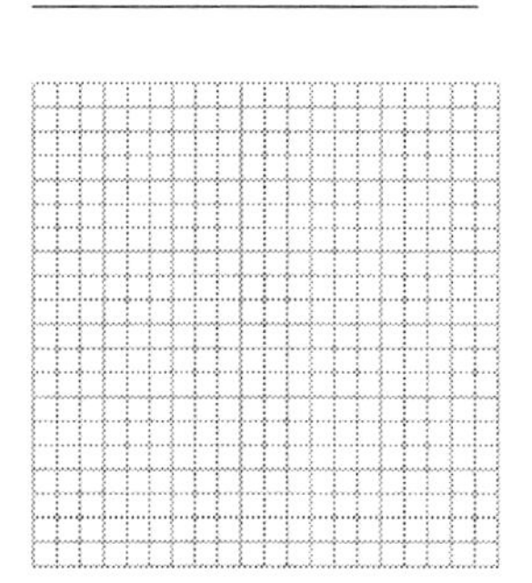

Objective 2 Graph cubic functions.

Graph each given cubic function, and state its domain and range.

7. $f(x)=(x+3)^3-8$

7.________________

____

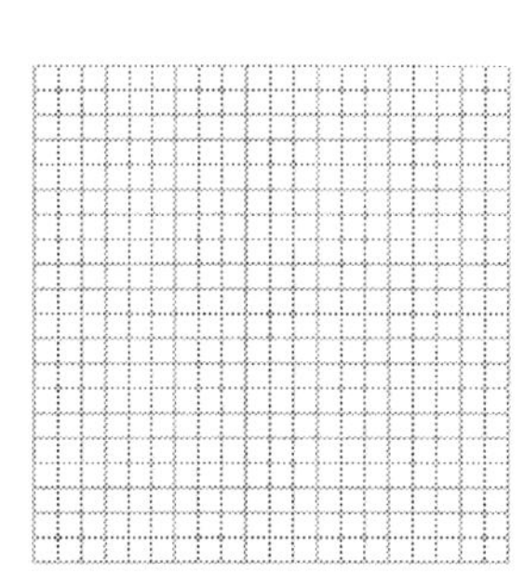

8. $f(x)=-(x+1)^3+1$

8.________________

____

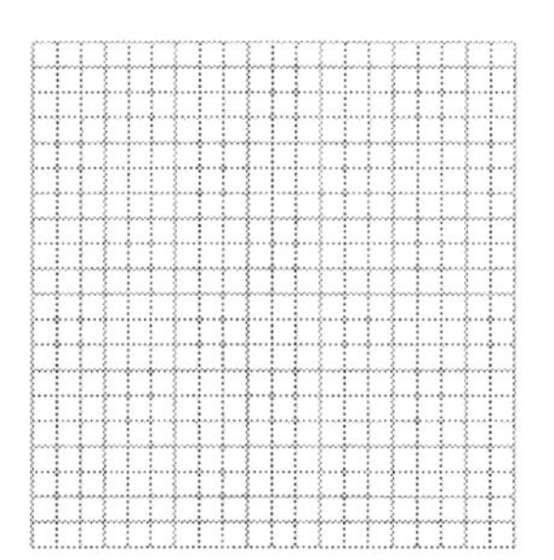

Name: Date:
Instructor: Section:

Objective 3 Determine a function from its graph.

Determine the function $f(x)$ that has been graphed. The function will be of the form $f(x) = a\sqrt{x-h} + k$ or $f(x) = a(x-h)^3 + k$. You may assume that $a = 1$ or $a = -1$.

9. **9.**________________

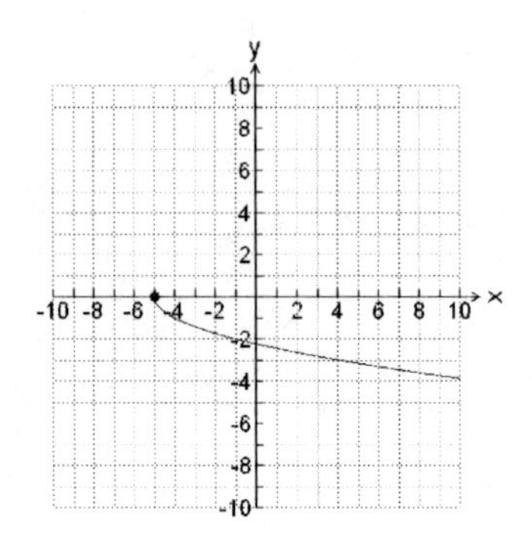

10. **10.**________________

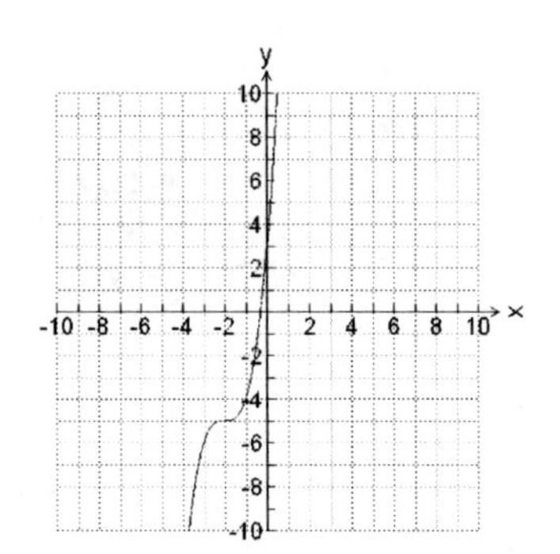

Name: Date:
Instructor: Section:

11. **a.** On the same set of axes, graph the functions $f(x)=x^3$, $g(x)=(x+3)^3$, and $h(x)=(x-4)^3$.

b. Using the results from part (a), explain how the graph of $f(x)=(x+22)^3$ would differ from the graph of the function $f(x)=x^3$. Also, explain how the graph of $f(x)=(x-67)^3$ would differ from the graph of the function $f(x)=x^3$.

11. a.

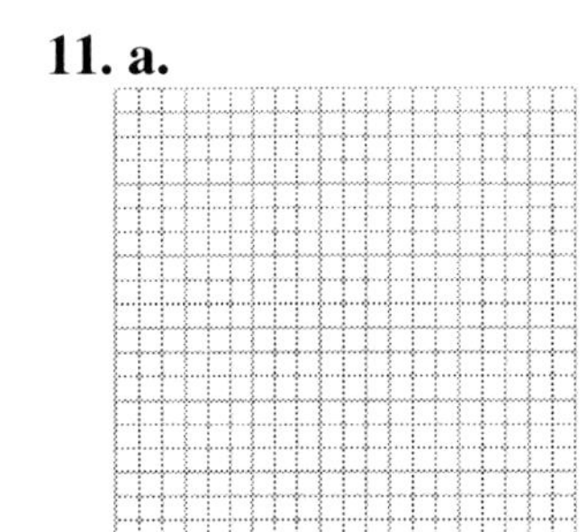

b.________________

12. **a.** On the same set of axes, graph the functions $f(x)=x^3$, $g(x)=x^3+2$, and $h(x)=x^3-8$.

b. Using the results from part (a), explain how the graph of $f(x)=x^3+15$ would differ from the graph of the function $f(x)=x^3$. Also, explain how the graph of $f(x)=x^3-31$ would differ from the graph of the function $f(x)=x^3$.

12. a.

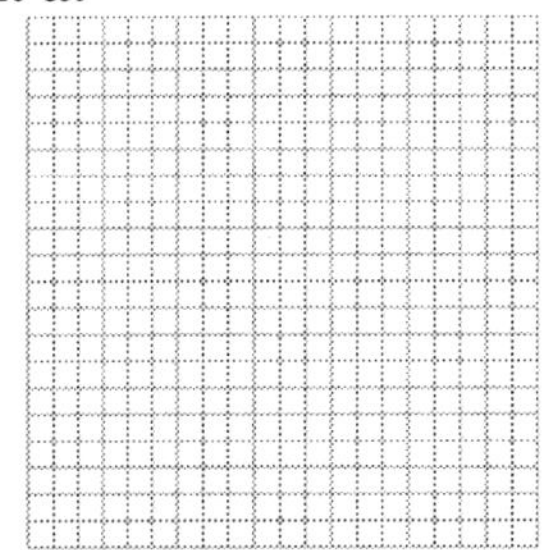

b.________________

Name: Date:
Instructor: Section:

13. On the same set of axes, graph the functions $f(x) = x^3$ and $g(x) = -x^3$. Explain, in general, how the graph of any cubic function is affected by placing a negative sign in front of the expression being cubed.

13.

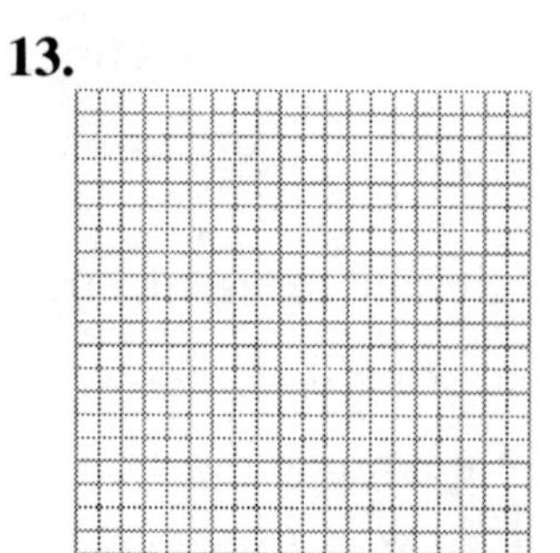

____

Name: Date:
Instructor: Section:

Chapter 8 LOGARITHMIC AND EXPONENTIAL FUNCTIONS

8.1 The Algebra of Functions

Learning Objectives

1 Find the sum, difference, product, and quotient functions for two functions $f(x)$ and $g(x)$.
2 Find the graph of the sum function or difference function from the graphs of two functions $f(x)$ and $g(x)$.
3 Solve applications involving the sum function or the difference function.
4 Find the composite function of two functions $f(x)$ and $g(x)$.
5 Find the domain of a composite function $(f \circ g)(x)$.

Key Terms

Use the most appropriate term from the given list to complete each statement in exercises 1-3.

$(f+g)(x)$ $(f \circ g)(x)$ $(f \cdot g)(x)$ $(g \circ f)(x)$

1. The composite function for $f(x)$ and $g(x)$ is denoted by __________________ or __________________.

2. The sum of the functions $f(x)$ and $g(x)$ is denoted by __________________.

3. The product of the functions $f(x)$ and $g(x)$ is denoted by __________________.

Objective 1 Find the sum, difference, product, and quotient functions for two functions f(x) and g(x).

For the given functions $f(x)$ and $g(x)$, find $(f+g)(x)$.

1. $f(x) = -x + 15$, $g(x) = 2x - 9$ **1.**__________________

For the given functions $f(x)$ and $g(x)$, find $(f-g)(x)$.

2. $f(x) = 3x + 10$, $g(x) = -x + 6$ **2.**__________________

3. $f(x) = 3x^2 - 7x - 20$, $g(x) = x^2 + 9x - 15$ **3.**__________________

Name: Date:
Instructor: Section:

For the given functions $f(x)$ *and* $g(x)$, *find* $(f \cdot g)(x)$.

4. $f(x) = x^2 - 8x + 24$, $g(x) = x - 9$ **4.**_______________

For the given functions $f(x)$ *and* $g(x)$, *find* $\left(\frac{f}{g}\right)(x)$.

5. $f(x) = x + 5$, $g(x) = x^2 + 13x + 40$ **5.**_______________

Let $f(x) = 5x - 13$ *and* $g(x) = x^2 - 7x - 10$. *Find the following.*

6. $(f + g)(-11)$ **6.**_______________

7. $(f - g)(-8)$ **7.**_______________

8. $(f \cdot g)(6)$ **8.**_______________

9. $\left(\frac{f}{g}\right)(-3)$ **9.**_______________

Find the unknown function $g(x)$ *that satisfies the given conditions.*

10. $f(x) = x + 6$, $(f \cdot g)(x) = x^2 - 7x - 78$ **10.**_______________

Name: Date:
Instructor: Section:

11. $f(x) = x^2 + 10x - 24$, $\left(\frac{f}{g}\right)(x) = x + 12$

11.________________

Objective 2 Find the graph of the sum function or difference function from the graphs of two functions *f*(*x*) and *g*(*x*).

Given the graphs of the functions $f(x)$ *and* $g(x)$, *graph the function* $(f+g)(x)$.

12.

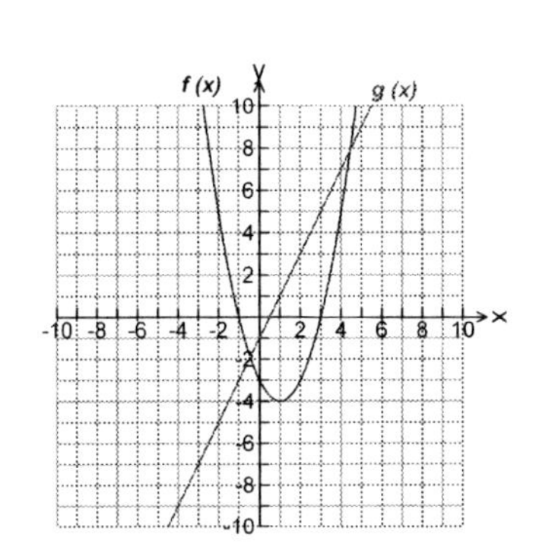

12.

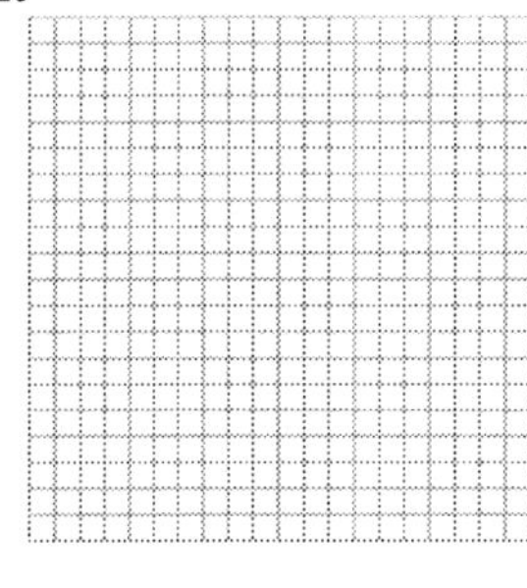

13.

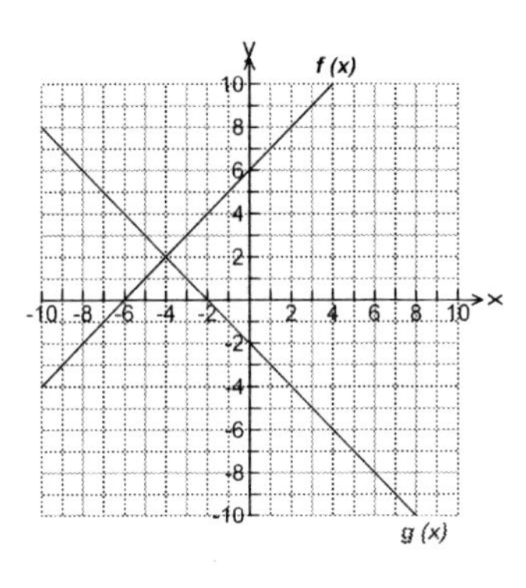

13.

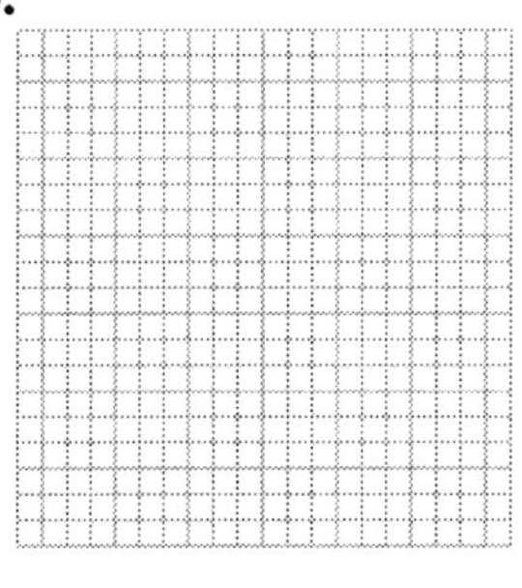

14.

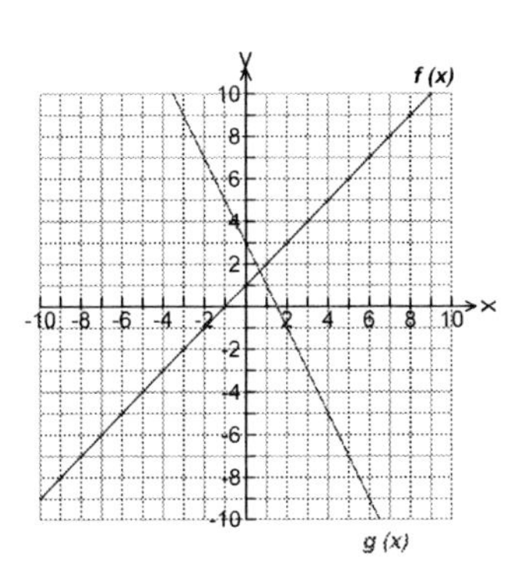

14.

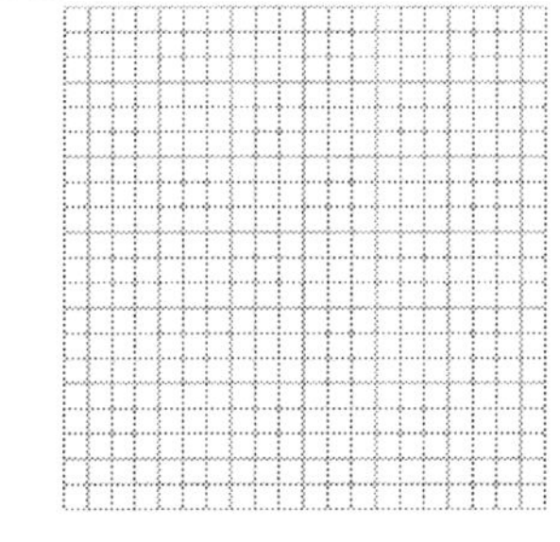

Name: Date:
Instructor: Section:

Given the graphs of the functions $f(x)$ *and* $g(x)$, *graph the function* $(f-g)(x)$.

15.

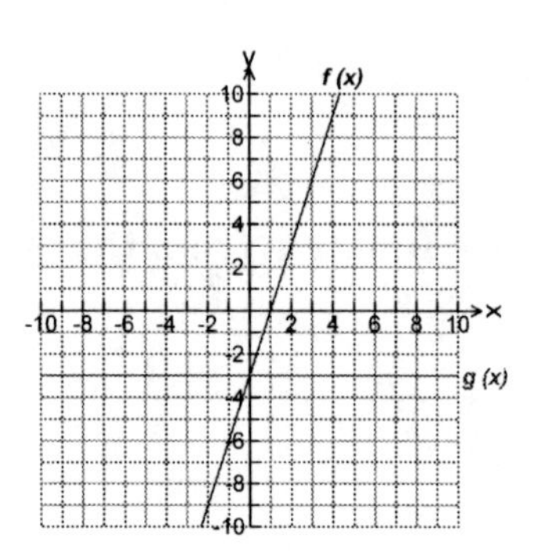

15.

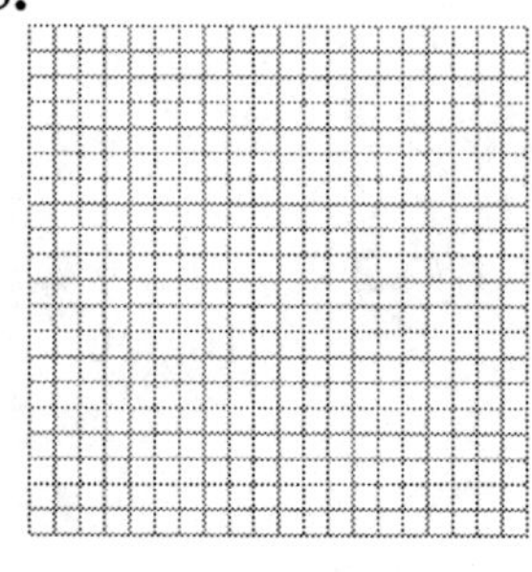

16.

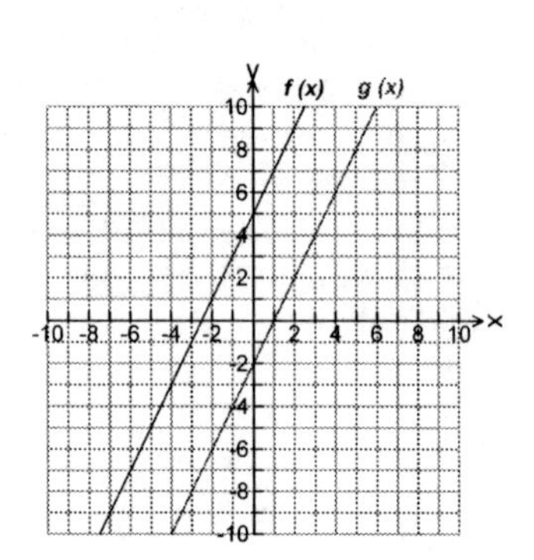

16.

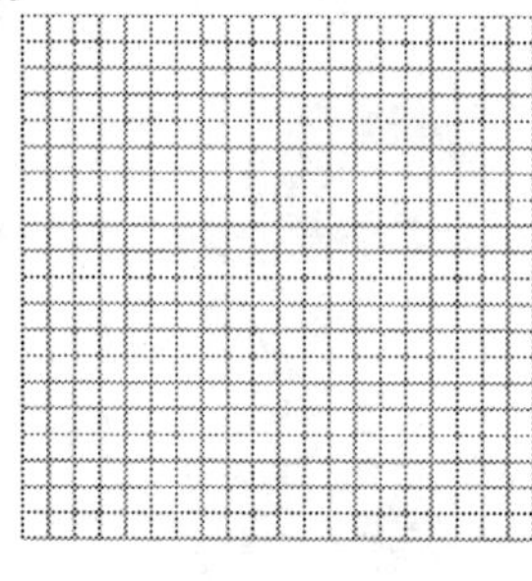

17.

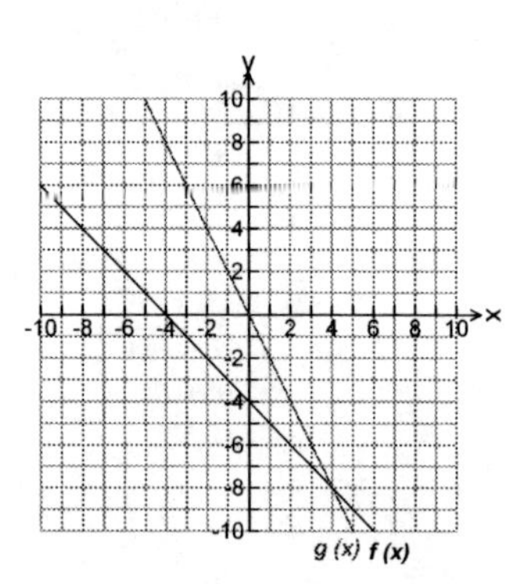

17.

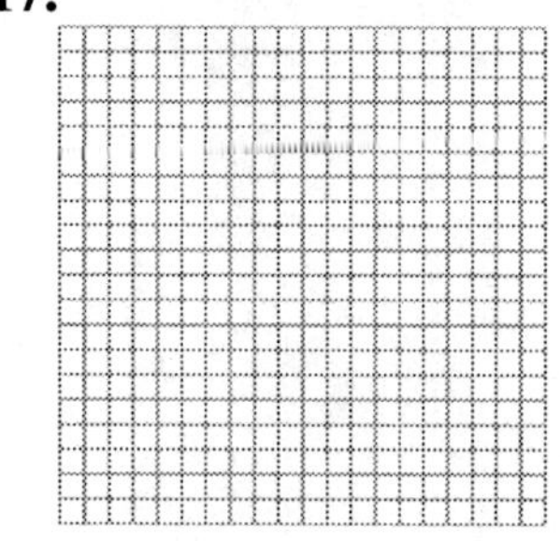

Name: Date:
Instructor: Section:

Objective 3 Solve applications involving the sum function or the difference function.

Solve.

18. The number of students enrolled at a public college in the United States in a particular year can be approximated by the function $f(x) = 111{,}916x + 9{,}528{,}000$, where x represents the number of years after 1990. The number of students enrolled at a private college in the United States in a particular year can be approximated by the function $g(x) = 43{,}973x + 2{,}591{,}000$, where again x represents the number of years after 1990. (*Source: U.S. Department of Education, National Center for Education Statistics*)

a. Find $(f+g)(25)$. Explain, in your own words, what this number represents.

b. Find $(f-g)(21)$. Explain, in your own words, what this number represents.

18.a. ________________

____

b. ________________

____

19. The number of births, in thousands, in the United States in a particular year can be approximated by the function $f(x) = 6x^2 - 72x + 4137$, where x represents the number of years after 1990. The number of deaths, in thousands, in the United States in a particular year can be approximated by the function $g(x) = 25x + 2159$, where again x represents the number of years after 1990. (*Source: National Center for Health Statistics, U.S. Department of Health and Human Services*)

a. Find $(f-g)(x)$. Explain, in your own words, what this function represents.

b. Find $(f-g)(40)$. Explain, in your own words, what this number represents.

19.a. ________________

____

b. ________________

____

Name: Date:
Instructor: Section:

Objective 4 Find the composite function of two functions $f(x)$ and $g(x)$.

Let $f(x)=x-5$ and $g(x)=3x+2$. Find the following.

20. $(f \circ g)(6)$ **20.**_______________

21. $(g \circ f)(-9)$ **21.**_______________

Objective 5 Find the domain of a composite function $(f \circ g)(x)$.

For the given functions $f(x)$ and $g(x)$, find $(f \circ g)(x)$ and state its domain.

22. $f(x)=\sqrt{x+6}$, $g(x)=x^2+7x+4$ **22.**_______________

____

23. $f(x)=\sqrt{x-36}$, $g(x)=x^2-x-36$ **23.**_______________

____

For the given function $g(x)$, find a function $f(x)$ such that $(f \circ g)(x)=x$.

24. $g(x)=x-9$ **24.**_______________

25. $g(x)=2x$ **25.**_______________

Name: Date:
Instructor: Section:

Chapter 8 LOGARITHMIC AND EXPONENTIAL FUNCTIONS

8.2 Inverse Functions

Learning Objectives

1 Determine whether a function is one to one.
2 Use the horizontal-line test to determine whether a function is one to one.
3 Understand inverse functions.
4 Determine whether two functions are inverse functions.
5 Find the inverse of a one-to-one function.
6 Find the inverse of a function from its graph.

Key Terms

Use the most appropriate term or phrase from the given list to complete each statement in exercises 1-3.

domain **range** **one to one** **inverse** **vertical** **horizontal**

1. A function is one to one if no ________________ line can intersect its graph at more than one point.

2. If $f(x)$ is different for each value in its ________________, then $f(x)$ is one to one.

3. The ________________ of a one-to-one function "undoes" what the function does.

Objective 1 Determine whether a function is one to one.

Determine whether the function $f(x)$ *is a one-to-one function.*

1. 1.________________

x	-8	-3	1	4	8
$f(x)$	8	3	-1	-4	-8

2. 2.________________

x	0	1	2	3	4
$f(x)$	6	9	12	9	6

Name: Date:
Instructor: Section:

Determine whether the function represented by the set of ordered pairs is a one-to-one function.

3. $\{(0, 0), (1, 1), (4, 2), (9, 3), (16, 4)\}$ **3.**______________

4. $\{(-7, 5), (-3, 5), (0, 5), (2, 5), (5, 5)\}$ **4.**______________

5. Is the function whose input is a former President of the United States and output is that President's party affiliation a one-to-one function? Explain your answer in your own words. **5.**______________

6. Is the function whose input is a state and output is that state's capital city a one-to-one function? Explain your answer in your own words. **6.**______________

7. Is the function whose input is a high school student and output is that student's first choice for a college to attend a one-to-one function? Explain your answer in your own words. **7.**______________

Objective 2 Use the horizontal-line test to determine whether a function is one to one.

Use the horizontal line test to determine whether the function is one-to-one.

8. **8.**______________

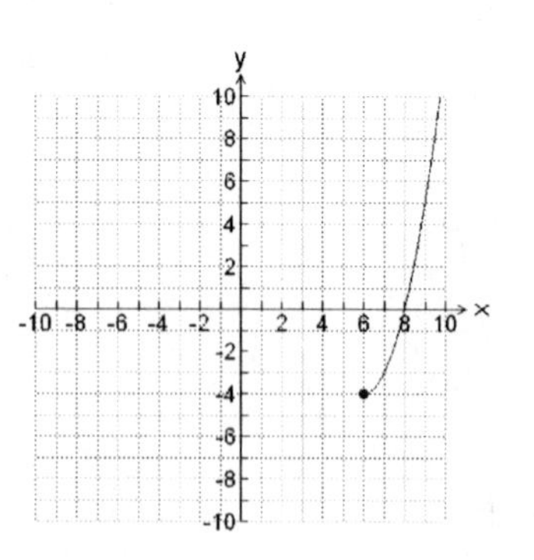

Name: Date:
Instructor: Section:

9.

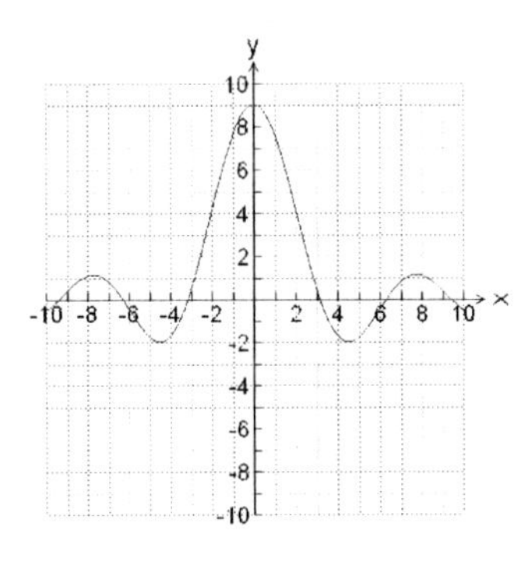

9.______________

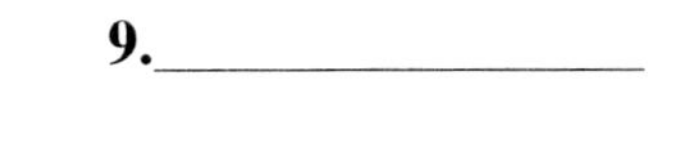

Objective 3 Understand inverse functions.
Objective 4 Determine whether two functions are inverse functions.

Determine whether the functions $f(x)$ *and* $g(x)$ *are inverse functions by showing that* $(f \circ g)(x) = x$ and $(g \circ f)(x) = x$.

10. $f(x) = 5x + 8,\ g(x) = \dfrac{x-8}{5}$

10.______________

11. $f(x) = \dfrac{x-10}{6},\ g(x) = 6x - 10$

11.______________

12. $f(x) = \dfrac{1}{x} + 4,\ g(x) = \dfrac{1}{x-4}$

12.______________

13. $f(x) = \dfrac{5}{x-9},\ g(x) = \dfrac{5+9x}{x}$

13.______________

Name: Date:
Instructor: Section:

14. $f(x) = -2x$, $g(x) = 2x$

14.________________

Objective 5 Find the inverse of a one-to-one function.

For the given function $f(x)$, find $f^{-1}(x)$.

15. $f(x) = x$

15.________________

16. $f(x) = -x$

16.________________

17. $f(x) = \dfrac{6}{2x-7}$

17.________________

18. $f(x) = \dfrac{-12x}{5x-4}$

18.________________

For the given function $f(x)$, find $f^{-1}(x)$. State the domain of $f^{-1}(x)$.

19. $f(x) = x^2 \ (x \geq 0)$

19.________________

Name: Date:
Instructor: Section:

20. $f(x)=(x-3)^2 \ (x\geq 3)$

20.________________

If the function represented by the set of ordered pairs is one-to-one, find its inverse.

21. $\{(4,\ 7),\ (5,\ 8),\ (7,\ 9),\ (11,\ 10),\ (19,\ 11)\}$

21.________________

22. $\{(-6,\ 1),\ (-4,\ -3),\ (-2,\ 1),\ (0,\ 13),\ (2,\ 33)\}$

22.________________

Objective 6 Find the inverse of a function from its graph.

For the given graph of a one-to-one function $f(x)$, graph its inverse function $f^{-1}(x)$.

23.

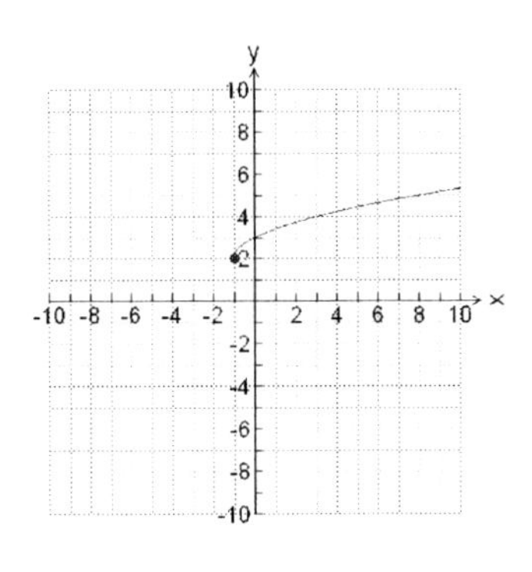

23.

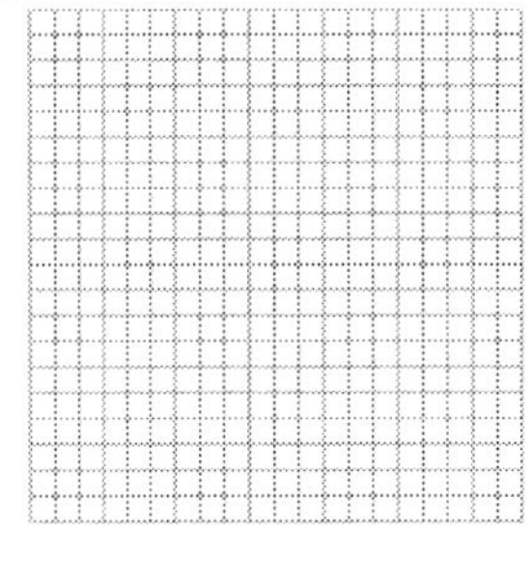

24.

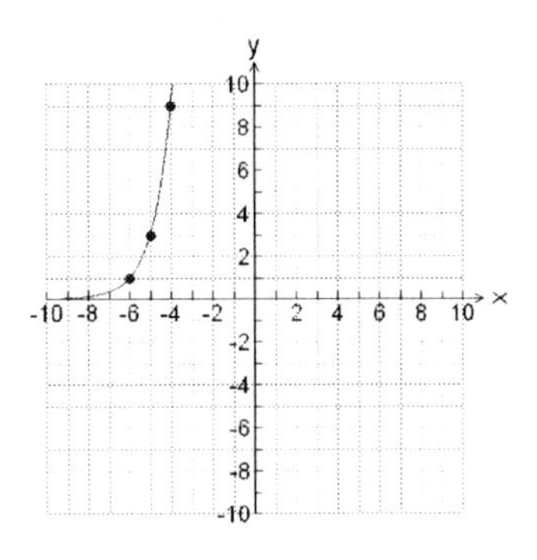

24.

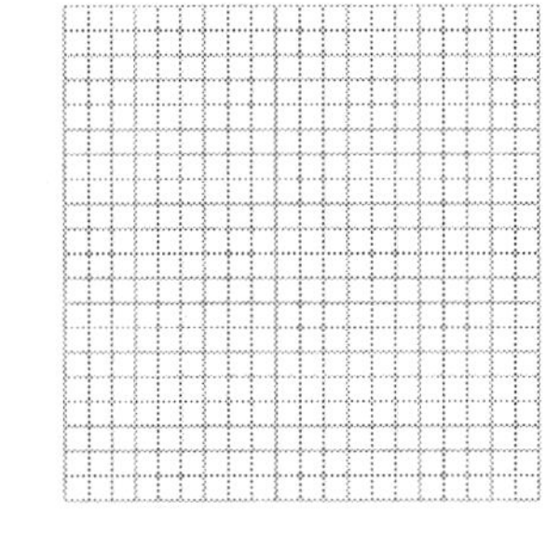

Name: Date:
Instructor: Section:

25.

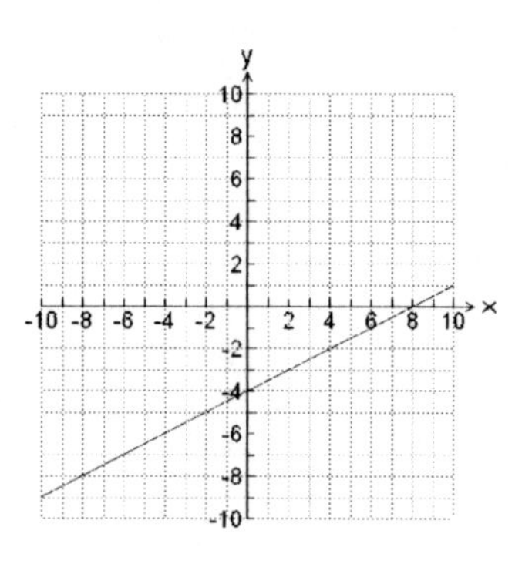

25.

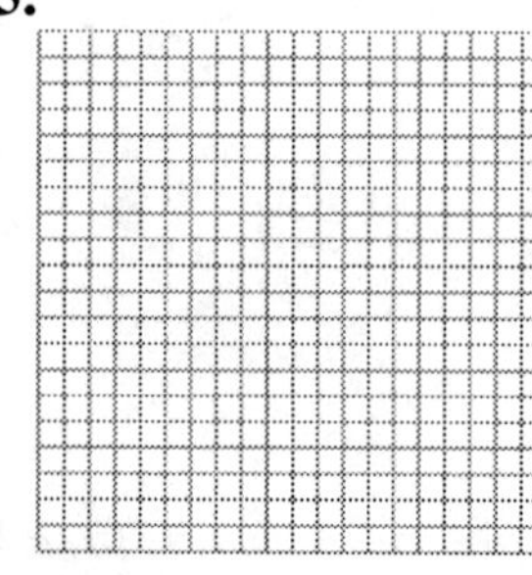

26.

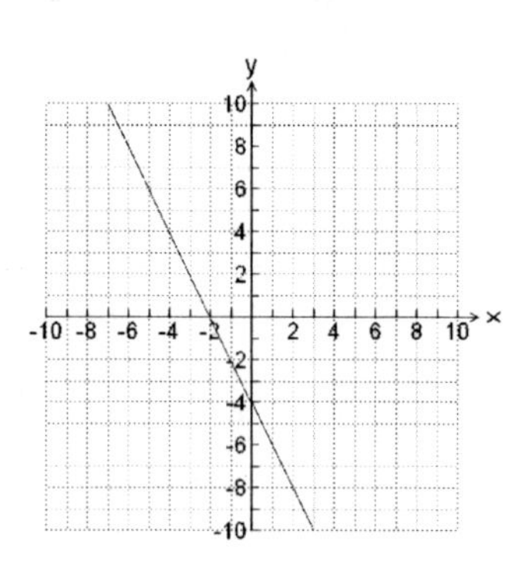

26.

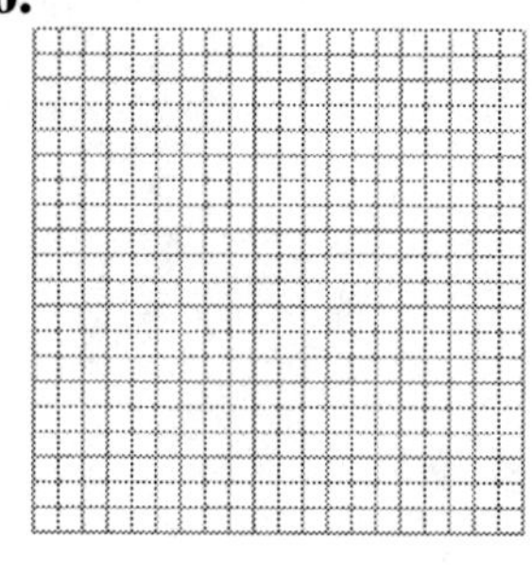

27.

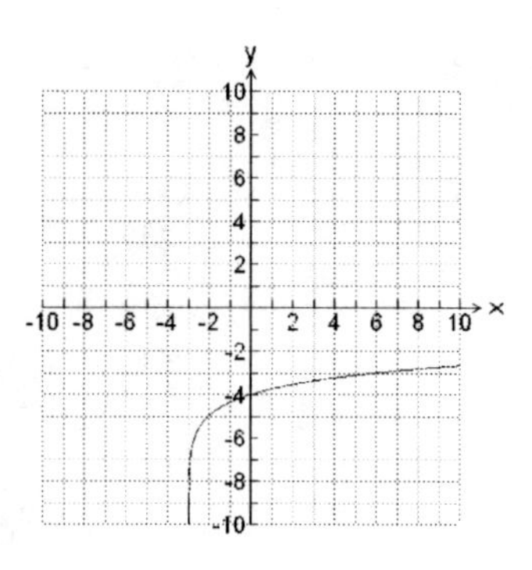

27.

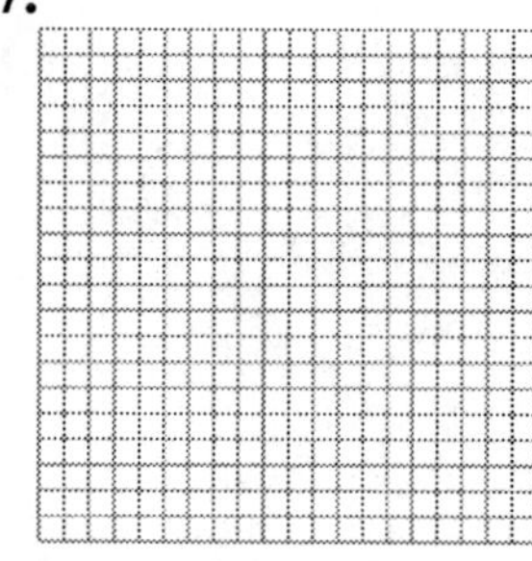

28.

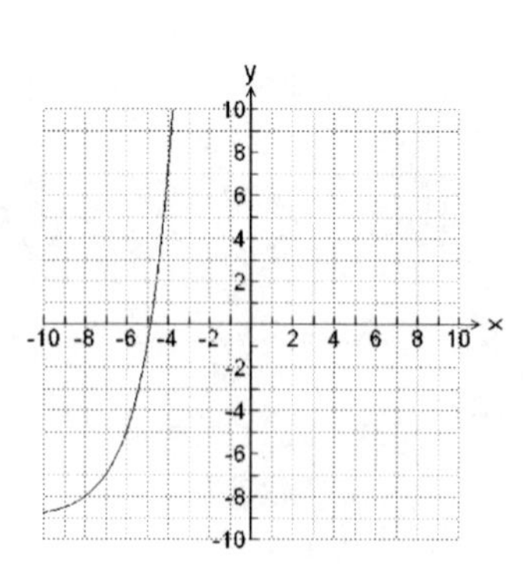

28.

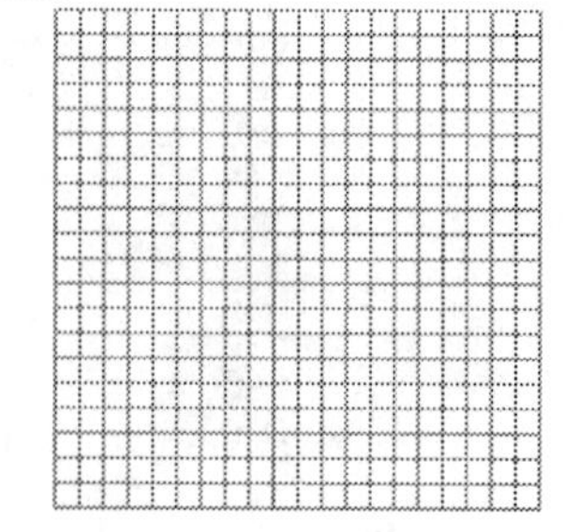

Name: Date:
Instructor: Section:

Graph a one-to-one function $f(x)$ that meets the given criteria.

29. $f(x)$ is a linear function, $f(7)=3$ and $f^{-1}(0)=-8$.

29.

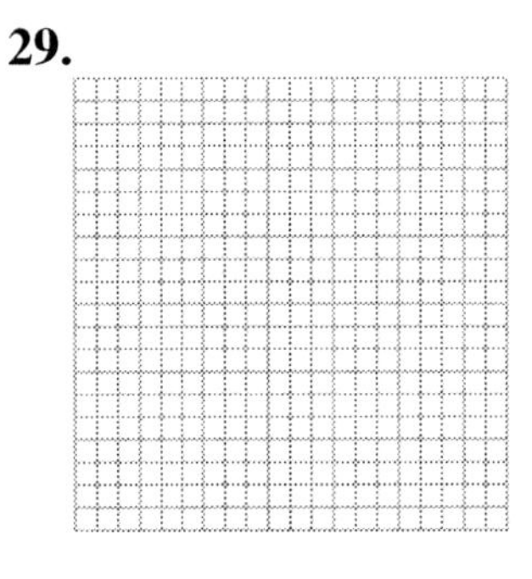

30. $f(x)$ is a linear function, $f^{-1}(-2)=-2$ and $f^{-1}(-6)=3$.

30.

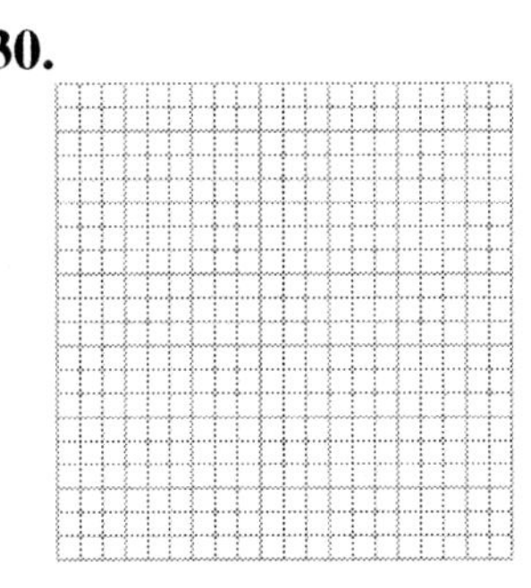

Name: Date:
Instructor: Section:

Chapter 8 LOGARITHMIC AND EXPONENTIAL FUNCTIONS

8.3 Exponential Functions

Learning Objectives
1. Define exponential functions.
2. Evaluate exponential functions.
3. Graph exponential functions.
4. Define the natural exponential function.
5. Solve exponential equations.
6. Use exponential functions in applications.

Key Terms

Use the most appropriate term or phrase from the given list to complete each statement in exercises 1-3.

base **exponent** **asymptote** **increasing** **decreasing**

1. A horizontal ______________________ is a line that the graph of a function approaches as it either moves to the left or to the right.

2. A function that can be written in the form $f(x) = b^x$ where $b > 0$ and $b \neq 1$ is called an exponential function with ______________________ b.

3. A function $f(x)$ is ______________________ if $f(x)$ increases as x increases.

Objective 1 Define exponential functions.

Let $f(x) = 2^x$. *Find the following.*

1. $f(-3)$ **1.** ________________

2. $f(3)$ **2.** ________________

Name: Date:
Instructor: Section:

Let $f(x)=\left(\frac{2}{5}\right)^x$. *Find the following.*

3. $f(-1)$ **3.**____________

4. $f(0)$ **4.**____________

Objective 2 Evaluate exponential functions.

Evaluate the given function..

5. $f(x)=\left(\frac{2}{3}\right)^{x+7}-2,\ f(-5)$ **5.**____________

6. $f(x)=\left(\frac{3}{4}\right)^{x-5}-16, f(4)$ **6.**____________

7. $f(x)=2^{x+7}+39,\ f(-10)$ **7.**____________

Complete the table for $f(x)$.

8. $f(x)=2^{x+2}$ **8.**____________

x	-4	-3	-2	-1	0
$f(x)$					

Name: Date:
Instructor: Section:

9. $f(x) = 3^x + 7$ **9.**______________

x	-2	-1	0	1	2
$f(x)$					

Objective 3 Graph exponential functions.

Graph $f(x)$. Label the horizontal asymptote. State the domain and range of the function.

10. $f(x) = 7^{-x} + 6$

10.______________

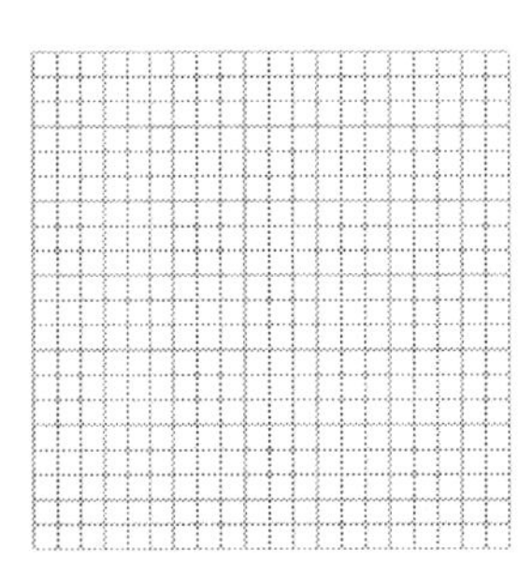

11. $f(x) = 10^{x+3} - 10$ **11.**______________

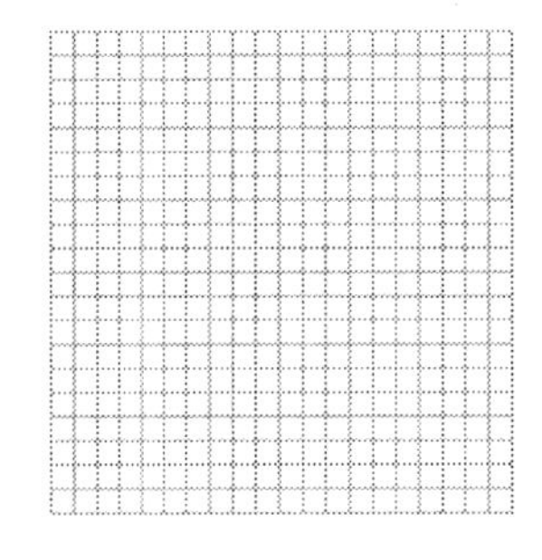

Objective 4 Define the natural exponential function.

Evaluate the given function. Round to the nearest thousandth.

12. $f(x) = e^{x^2-6} + 7,\ f(3)$ **12.**______________

Name: Date:
Instructor: Section:

13. $f(x)=e^{x^2+6x}-8,\ f(-7)$

13.________________

Objective 5 Solve exponential equations.

Solve.

14. $\left(\frac{1}{2}\right)^x=\frac{1}{8}$

14.________________

15. $\left(\frac{1}{5}\right)^x=\frac{1}{125}$

15.________________

16. $\left(\frac{3}{4}\right)^x=\frac{27}{64}$

16.________________

17. $\left(\frac{5}{8}\right)^x=\frac{25}{64}$

17.________________

Name: Date:
Instructor: Section:

18. $\left(\frac{7}{2}\right)^x = \frac{8}{343}$

18._______________

19. $\left(\frac{2}{5}\right)^x = \frac{3125}{32}$

19._______________

20. $3^{x^2+8x+14} = 9$

20._______________

21. $4^{x^2+10x-21} = 64$

21._______________

Objective 6 Use exponential functions in applications.

Solve.

22. Terry deposited \$22,500 in an account that pays 3% annual interest, compounded monthly. The balance after t years is given by the function $f(t) = 22,500(1.0025)^{12t}$. What will the balance of this account be after 15 years?

22._______________

Name: Date:
Instructor: Section:

23. Patti deposited \$100,000 in an account that pays 6% annual interest, compounded monthly. The balance after t years is given by the function $f(t)=100,000(1.005)^{12t}$. What will the balance of this account be after 30 years?

23.________________

24. The number of *Staphylococcus aureus* cells in a Petri dish after x minutes can be approximated by the function $f(x)=25,000\cdot e^{0.023105x}$, if there were 25,000 cells present initially. How many cells will be present after 90 minutes?

24.________________

25. The number of *Lactobacillus acidophilus* cells in a Petri dish after x minutes can be approximated by the function $f(x)=5000\cdot e^{0.008664x}$, if there were 5000 cells present initially. How many cells will be present after 150 minutes?

25.________________

Name: Date:
Instructor: Section:

26. The cost of health insurance has been rising exponentially. The amount spent on health insurance by state and local government for its full-time employees in a particular year can be approximated by the function $f(t)=3016(1.07)^t$, where t represents the number of years after 1993. Use this function to approximate the amount that state and local governments will spend on health insurance for a full-time employee in the year 2015.

26.________________

27. The population of the United States in a particular year, in millions of people, can be approximated by the function $f(t)=201.05(1.00996)^t$, where t represents the number of years after 1968. Assuming that this rate of growth continues, use this function to predict the U.S. population in the year 2018.

27.________________

Name: Date:
Instructor: Section:

Plot the given ordered pairs $(x, f(x))$. Based on the graph, do you feel that $f(x)$ is an exponential function? Explain your reasoning.

28. $(-8, -7), (-7, -6), (-5, -5), (-1, -4), (7, -3)$

28.________________

29. $(0, -1), (1, 1), (2, 7), (3, 8), (5, 9), (9, 10)$

29.________________

Name: Date:
Instructor: Section:

Chapter 8 LOGARITHMIC AND EXPONENTIAL FUNCTIONS

8.4 Logarithmic Functions

Learning Objectives	
1	Define logarithms and logarithmic functions.
2	Evaluate logarithms and logarithmic functions.
3	Define the common logarithm and natural logarithm.
4	Convert back and forth between exponential form and logarithmic form.
5	Solve logarithmic equations.
6	Graph logarithmic functions.
7	Use logarithmic functions in applications.

Key Terms

Use the most appropriate term or phrase from the given list to complete each statement in exercises 1-2.

decimal **argument** **logarithmic** **common** **natural**

1. The ___________________ logarithm is a logarithm whose base is 10.

2. For any positive real number b such that $b \neq 1$ a function of the form $f(x) = \log_b x$ is called a ___________________ function. The number x is called the ___________________ of the logarithm.

Objective 1 Define logarithms and logarithmic functions.
Objective 2 Evaluate logarithms and logarithmic functions.

Evaluate.

1. $\log_{1/2}\left(\frac{1}{32}\right)$ **1.**_______________

2. $\log_{2/3}\left(\frac{64}{729}\right)$ **2.**_______________

3. $\log_{1/3} 243$ **3.**_______________

Name: Date:
Instructor: Section:

4. $\log_{1/5} 25$ **4.**________________

5. $\log_6 \left(6^8\right)$ **5.**________________

6. $\log_9 \left(9^{16}\right)$ **6.**________________

Evaluate the given function.

7. $f(x) = \log_7 x$ **7.a.**________________

a. $f(49)$ **b.**________________

b. $f\left(\frac{1}{7}\right)$

8. $f(x) = \log_4 x$ **8.a.**________________

a. $f(256)$ **b.**________________

b. $f\left(\frac{1}{64}\right)$

Objective 3 Define the common logarithm and natural logarithm.

Evaluate. Round to the nearest thousandth if necessary.

9. $\log 345$ **9.**________________

10. $\log 12,408$ **10.**________________

11. $\log 31.9$ **11.**________________

Name: Date:
Instructor: Section:

12. $\log 0.317$ **12.**________________

13. $\ln 2$ **13.**________________

14. $\ln 5$ **14.**________________

15. $\ln 3.104$ **15.**________________

16. $\ln 0.039$ **16.**________________

Objective 4 Convert back and forth between exponential form and logarithmic form.

Rewrite in logarithmic form.

17. $3^{12} = 531{,}441$ **17.**________________

18. $7^{x} = 343$ **18.**________________

19. $3^{x} + 6 = 20$ **19.**________________

20. $5^{x} - 11 = 13$ **20.**________________

Rewrite in exponential form.

21. $\log_{2} 22 = x$ **21.**________________

Name: Date:
Instructor: Section:

22. $\log_5 625 = 4$ **22.**________________

23. $\log_{0.3} 307 = x$ **23.**________________

24. $\ln(x+8) = 2$ **24.**________________

Objective 5 Solve logarithmic equations.

Solve. Round to the nearest thousandth if necessary.

25. $\log_5 x = 1.4$ **25.**________________

26. $\log_9 x = 2.5$ **26.**________________

27. $\log x = -1.95$ **27.**________________

28. $\ln x = 2.03$ **28.**________________

29. Let $f(x) = \log_2 x$. Solve $f(x) = -5$. **29.**________________

Name: Date:
Instructor: Section:

30. $\log_8 128 = x$

30.________________

Objective 6 Graph logarithmic functions.

Use the graph of the given exponential function $f(x)$ to graph its inverse logarithmic function $f^{-1}(x)$.

31.

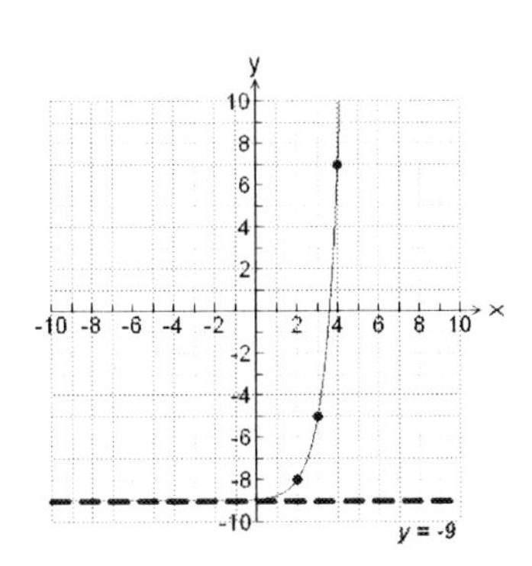

31.

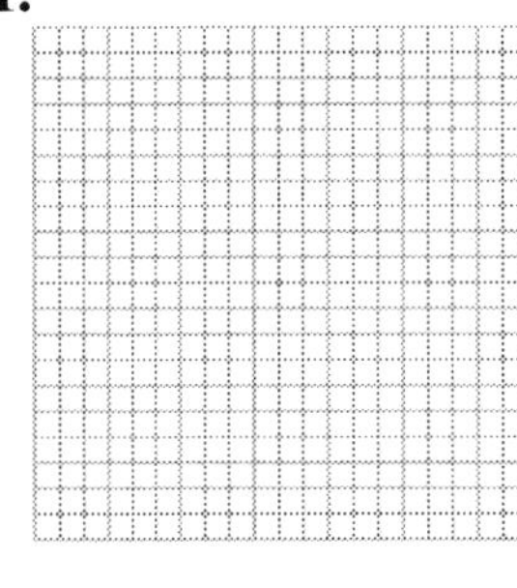

32.

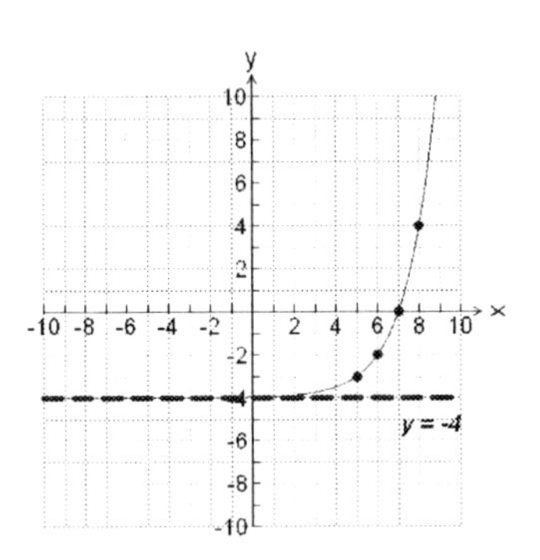

32.

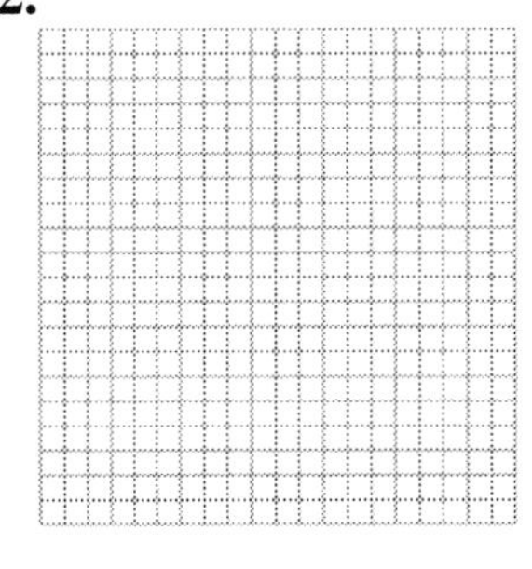

Objective 7 Use logarithmic functions in applications.

Solve.

33. The percent of U.S. drivers of passenger cars who wear their seatbelts in a particular year can be approximated by the function $f(x) = 64.9 + 6.6 \ln x$, where x represents the number of years after 1995. Use this function to predict what percent of U.S. drivers of passenger cars will wear their seatbelts in the year 2015.

33.________________

Name: Date:
Instructor: Section:

34. The number of automobile fatalities in the U.S. in a particular year can be approximated by the function $f(x) = 39,485 + 1221 \ln x$, where x represents the number of years after 1991. Use this function to predict the number of automobile fatalities in the U.S. in the year 2009.

34._______________

Use the formula $R = \log I$, where R is the magnitude of an earthquake on the Richter scale whose shock wave is I times larger than the smallest measurable shock wave that is recordable on a seismograph.

35. If an earthquake had a magnitude of 6.2 on the Richter scale, how many times larger was the shock wave created by this earthquake than the smallest measurable shock wave?

35._______________

36. If an earthquake had a magnitude of 5.4 on the Richter scale, how many times larger was the shock wave created by this earthquake than the smallest measurable shock wave?

36._______________

Name: Date:
Instructor: Section:

Chapter 8 LOGARITHMIC AND EXPONENTIAL FUNCTIONS

8.5 Properties of Logarithms

Learning Objectives
1 Use the product rule for logarithms.
2 Use the quotient rule for logarithms.
3 Use the power rule for logarithms.
4 Use additional properties of logarithms.
5 Use properties to rewrite two or more logarithmic expressions as a single logarithmic expression.
6 Use properties to rewrite a single logarithmic expression as a sum or difference of logarithmic expressions whose arguments have an exponent of 1.
7 Use the change-of-base formula for logarithms.

Objective 1 Use the product rule for logarithms.

Rewrite as a single logarithm, using the product rule for logarithms. Simplify if possible. Assume all variables represent positive real numbers.

1. $\log_c F + \log_c Y$ **1.**________________

2. $\log 62 + \log 31$ **2.**________________

Rewrite as the sum of two or more logarithms, using the product rule for logarithms. Simplify if possible. Assume all variables represent positive real numbers.

3. $\log_8 3w$ **3.**________________

4. $\log_7 (49 \cdot 15)$ **4.**________________

Objective 2 Use the quotient rule for logarithms.

Rewrite as a single logarithm, using the quotient rule for logarithms. Simplify if possible. Assume all variables represent positive real numbers.

5. $\log_6 56 - \log_6 8$ **5.**________________

Name: Date:
Instructor: Section:

6. $\log 45 - \log y$ **6.**______________

Rewrite in terms of two or more logarithms, using the quotient and product rules for logarithms. Simplify if possible. Assume all variables represent positive real numbers.

7. $\log_5\left(\frac{14x}{yz}\right)$ **7.**______________

8. $\log_2\left(\frac{16z}{4w}\right)$ **8.**______________

Objective 3 Use the power rule for logarithms.

Rewrite, using the power and product rules. Simplify if possible. Assume all variables represent positive real numbers.

9. $\log_9 s^{14}$ **9.**______________

10. $\log\left(x^7 y^3\right)$ **10.**______________

Rewrite, using the power rule. Assume all variables represent positive real numbers.

11. $2\log_6 m$ **11.**______________

12. $\frac{1}{5}\log a$ **12.**______________

Objective 4 Use additional properties of logarithms.

Simplify.

13. $\ln e^{13}$ **13.**______________

Name: Date:
Instructor: Section:

14. $\log 10^6$ 14.________________

15. $9^{\log_9 x}$ 15.________________

16. $e^{\ln 45}$ 16.________________

Objective 5 Use properties to rewrite two or more logarithmic expressions as a single logarithmic expression.

Rewrite as a single logarithm. Assume that all variables represent positive real numbers. Simplify if possible.

17. $4\log_2 3 + 3\log_2 5 + 2\log_2 9$ 17.________________

18. $5\ln 2 + 2\ln 5 - 3\ln 4$ 18.________________

19. $\log_4(x-3) + \log_4(x-7) - 3\log_4 8$ 19.________________

20. $\log_2(x^2 - 7x + 10) - \log_2(x^2 + 7x - 18)$ 20.________________

21. $-\log_3 5 - \log_3 6$ 21.________________

22. $-\log 8 - \log 6 - \log x$ 22.________________

23. $-3\log_5 2 - \log_5 17$ 23.________________

Name: Date:
Instructor: Section:

24. $-2\log_3 7 - 6\log_3 2$ **24.**_______________

25. $\log_6 756 - \log_6 21$ **25.**_______________

26. $8\ln x + \frac{1}{5}\ln y$ **26.**_______________

Objective 6 Use properties to rewrite a single logarithmic expression as a sum or difference of logarithmic expressions whose arguments have an exponent of 1.

Rewrite as the sum or difference of logarithmic expressions whose arguments have an exponent of 1. Simplify, if possible. Assume that all variables represent positive real numbers.

27. $\ln\left(\frac{ab}{c}\right)$ **27.**_______________

28. $\log\left(x^5\sqrt{y}\right)$ **28.**_______________

29. $\log_3 729a^{12}$ **29.**_______________

Name: Date:
Instructor: Section:

Objective 7 Use the change-of-base formula for logarithms.

Suppose for some base $b > 0$ $(b \neq 1)$ *that* $\log_b 2 = A$, $\log_b 3 = B$, $\log_b 5 = C$, *and* $\log_b 7 = D$. *Express the following logarithms in terms of A, B, C, or D.*
(Hint: Rewrite the logarithm in terms of $\log_b 2$, $\log_b 3$, $\log_b 5$, *or* $\log_b 7$.*)*

30. $\log_b\left(\frac{2}{5}\right)$ **30.**________________

31. $\log_b\left(\frac{7}{3}\right)$ **31.**________________

32. $\log_b 56$ **32.**________________

33. $\log_b 75$ **33.**________________

Expand. Simplify, if possible. Assume that all variables represent positive real numbers.

34. $\log\left(\frac{1000\sqrt[5]{x}}{yz^5}\right)$ **34.**________________

35. $5\log_2 8x^3y^4$ **35.**________________

36. $4\log_3\left(\frac{243a^{10}b^7}{c^6}\right)$ **36.**________________

Name: Date:
Instructor: Section:

Evaluate using the change of base formula. Round to the nearest thousandth.

37. $\log_{50} 0.6$ **37.**________________

38. $\log_{90} 0.83$ **38.**________________

Find the missing number. Round to the nearest thousandth. (Hint: Rewrite in logarithmic form and use the change-of-base formula.)

39. $8^{?} = 6$ **39.**________________

40. $13^{?} = 0.49$ **40.**________________

Name: Date:
Instructor: Section:

Chapter 8 LOGARITHMIC AND EXPONENTIAL FUNCTIONS

8.6 Exponential and Logarithmic Equations

Learning Objectives
1. Solve an exponential equation in which both sides have the same base.
2. Solve an exponential equation by using logarithms.
3. Solve a logarithmic equation in which both sides have logarithms with the same base.
4. Solve a logarithmic equation by converting it to exponential form.
5. Use properties of logarithms to solve a logarithmic equation.
6. Find the inverse function for exponential and logarithmic functions.

Objective 1 Solve an exponential equation in which both sides have the same base.

Solve.

1. $2^{x+3} = \frac{1}{16}$ **1.**________________

2. $3^{2x-7} = 243$ **2.**________________

Objective 2 Solve an exponential equation by using logarithms.

Solve. Round to the nearest thousandth.

3. $e^{10x-3} = 145$ **3.**________________

4. $3^{2x+6} = 2.35$ **4.**________________

5. $10^{-x} = 46$ **5.**________________

Name: Date:
Instructor: Section:

6. $2^{-x} = 75$ **6.**________________

Objective 3 Solve a logarithmic equation in which both sides have logarithms with the same base.

Solve.

7. $\log(3x+16) = \log(7x+24)$ **7.**________________

8. $\log_{11}(x+10) = \log_{11}(-2x-11)$ **8.**________________

9. $\log_2(5x-19) = \log_2(3x+7)$ **9.**________________

10. $\ln(x-8) = \ln(3x+2)$ **10.**________________

Objective 4 Solve a logarithmic equation by converting it to exponential form.

Solve. Round to the nearest thousandth.

11. $\log(2x-9) - 13 = -11$ **11.**________________

12. $\log_4(3x+7) + 10 = 13$ **12.**________________

Name: Date:
Instructor: Section:

13. $2\ln(4x+7)+5=9$ **13.**________________

14. $5\log(6x-2)+21=36$ **14.**________________

15. $\log_2(x+6)=7$ **15.**________________

16. $\log_4(3x-11)+10=12$ **16.**________________

Objective 5 Use properties of logarithms to solve a logarithmic equation.

Solve.

17. $\log(x-28)+\log(x+20)=2$ **17.**________________

18. $\log_5 x+\log_5(6x-1)=0$ **18.**________________

19. Let $f(x)=\log_7(4x-3)-10$. Solve $f(x)=-8$. **19.**________________

Name: Date:
Instructor: Section:

20. Let $f(x) = \log_2(x^2 - 8x - 16)$. Solve $f(x) = 5$. **20.**________________

21. Let $f(x) = e^{2x-9} + 3$. Solve $f(x) = 7$. Round to the nearest thousandth. **21.**________________

22. Let $f(x) = \ln(x-3) - 9$. Solve $f(x) = -2$. Round to the nearest thousandth. **22.**________________

Objective 6 Find the inverse function for exponential and logarithmic functions.

For the given function $f(x)$, find its inverse function $f^{-1}(x)$.

23. $f(x) = e^{x+9} - 2$ **23.**________________

24. $f(x) = 2^{x-7} - 3$ **24.**________________

25. $f(x) = \log(x+9) - 8$ **25.**________________

26. $f(x) = \log_3(x+4) + 6$ **26.**________________

Name: Date:
Instructor: Section:

Chapter 8 LOGARITHMIC AND EXPONENTIAL FUNCTIONS

8.7 Applications of Exponential and Logarithmic Functions

Learning Objectives
1. Solve compound interest problems.
2. Solve exponential growth problems.
3. Solve exponential decay problems.
4. Solve applications involving exponential functions.
5. Solve applications involving logarithmic functions.

Objective 1 Solve compound interest problems.

Solve.

1. One account pays 9% annual interest, compounded quarterly, while a second account pays 9.6% annual interest, compounded monthly. If $7000 is deposited in each account, which will be the first account to produce a balance of $10,000? How long will it take?

1.________________

2. Which account has the shortest doubling time: 8% annual interest, compounded daily, or 8.4% annual interest, compounded quarterly? Explain your reasoning fully. Use 1 year = 365 days.

2.________________

3. What is the doubling time of an account that pays 6.9% annual interest, compounded continually?

3.________________

Name: Date:
Instructor: Section:

4. In an account that pays interest that is compounded continually, an initial deposit of \$5000 has grown to \$11,125 in 10 years. What is the annual interest rate on this account?

4.______________

Objective 2 Solve exponential growth problems.

Solve.

5. Between the years 1980 and 1995, a baseball card rose in value from \$10 to \$30. If the value of the card continues to grow exponentially at the same rate, in what year will the card be worth \$50?

5.______________

6. The U.S. gross movie box office receipts grew from \$6.4 billion in 1997 to \$8.4 billion in 2001. If these receipts continue to grow exponentially at this rate, when will the gross receipts reach \$25 billion?

6.______________

Objective 3 Solve exponential decay problems.

Solve.

7. In 1996, the amount of waste discarded was 17.6 million tons, but by 1998 the amount declined to 15.6 million tons. Assume the amount of discarded waste is decreasing according to the exponential decay model. Estimate the amount of waste discarded in 2008. Round to the nearest tenth.

7.______________

Name: Date:
Instructor: Section:

8. The amount of carbon-14 present in animal bones after t years is given by $y = y_0 e^{-0.00012t}$. A bone contains 29% of its carbon-14. How old is the bone? Round to the nearest year.

8.________________

Objective 4 Solve applications involving exponential functions.

Solve.

9. In the United States, the average tuition and fees charged by four-year public colleges can be described by the function $f(x) = 2398 \cdot 1.05^x$, where x represents the number of years after 1992. Use this function to predict the average tuition and fees that four-year public colleges will charge in 2009.

9.________________

10. The amount, in billions of dollars, spent in the U.S. on herbicides, insecticides, and other pesticides in a particular year can be described by the function $f(x) = 5.49 \cdot 1.035^x$, where x represents the number of years after 1979. Use this function to predict how much will be spent in the U.S. on these products in the year 2015.

10.________________

Name: Date:
Instructor: Section:

11. The number of doctoral degrees conferred to women has been growing exponentially. The number of doctoral degrees awarded to women can be described by the function $f(x) = 8083 \cdot 1.039^x$, where x represents the number of years after 1975. Use this function to predict when 50,000 doctoral degrees will be conferred to women in a single year.

11.________________

12. The total amount of scholarship and fellowship awards given by colleges has been growing exponentially. The amount, in millions of dollars, given in a particular year can be described by the function $f(x) = 2.6 \cdot 1.127^x$, where x represents the number of years after 1981. Use this function to determine when the total amount of scholarship and fellowship awards given in a single year will reach $50 million.

12.________________

Objective 5 Solve applications involving logarithmic functions.

Solve. Use the formula $\text{pH} = -\log\left[\text{H}^+\right]$ *where* $\left[\text{H}^+\right]$ *is the concentration of hydrogen ions in moles per liter.*

13. The concentration of hydrogen ions $[\text{H}^+]$ in egg whites is 1×10^{-6} moles/liter. Find the pH of egg whites. Is an egg white an acid or a base?

13.________________

Name: Date:
Instructor: Section:

14. The concentration of hydrogen ions $[\mathrm{H}^+]$ in baking soda is 3.98×10^{-9} moles/liter. Find the pH of baking soda. Is baking soda an acid or a base?

14.________________

15. The pH of blood is 7.4. Find the concentration of hydrogen ions in moles per liter.

15.________________

16. The pH of household lye is 13.5. Find the concentration of hydrogen ions in moles per liter.

16.________________

Solve. Use the formula $L = 10\log\left(\frac{I}{10^{-12}}\right)$ *where I is the intensity of the sound in watts per square meter.*

17. The volume of a crowd at a football game can reach 117 db. Find the intensity of a sound with this volume.

17.________________

Name: Date:
Instructor: Section:

18. The volume of a popping balloon is 157 db. Find the intensity of a sound with this volume.

18.________________

19. The percent of all U.S. households that have cable TV in a particular year can be described by the function $f(x) = 56.3 + 6.25 \ln x$, where x represents the number of years after 1989. Use this function to predict the percent of all U.S. households that have cable TV in the year 2007.

19.________________

20. The percent of U.S. public schools that have Internet access in a particular year can be described by the function $f(x) = 31.4 + 33.77 \ln x$, where x represents the number of years after 1993. Use this function to estimate what percent of US public schools had Internet access in the year 1998.

20.________________

Name: Date:
Instructor: Section:

Chapter 8 LOGARITHMIC AND EXPONENTIAL FUNCTIONS

8.8 Graphing Exponential and Logarithmic Functions

Learning Objectives
1 Graph exponential functions.
2 Graph logarithmic functions.

Key Terms
Use the most appropriate term or phrase from the given list to complete each statement in exercises 1-3.

logarithmic **exponential** **restricted** **all real numbers**

1. The domain of an exponential function is __________________.

2. The graph of a __________________ function has a vertical asymptote.

3. The graph of every __________________ function has a y-intercept.

Objective 1 Graph exponential functions.
Objective 2 Graph logarithmic functions.

Find all intercepts for the given function. Find exact values, and then round to the nearest tenth if necessary.

1. $f(x) = \log(x+5)+3$ **1.**________________

2. $f(x) = \ln(x-3)$ **2.**________________

3. $f(x) = 4^{x+2}+18$ **3.**________________

4. $f(x) = 2^{x+1}+9$ **4.**________________

Name: Date:
Instructor: Section:

Determine the equation of the horizontal or vertical asymptote for the graph of this function and state the domain and range of this function.

5. $f(x) = e^x + 16$ **5.**________________

6. $f(x) = \left(\frac{1}{3}\right)^{x+3} - \frac{1}{9}$ **6.**________________

7. $f(x) = \ln(x+7) - 300$ **7.**________________

8. $f(x) = \ln(x-10) - \frac{37}{2}$ **8.**________________

Graph. Label any intercepts and asymptotes. State the domain and range of the function.

9. $f(x) = \log_2(x-4) - 3$ **9.**________________

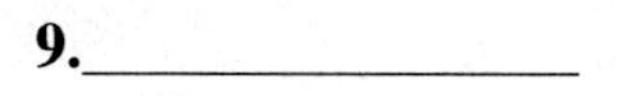

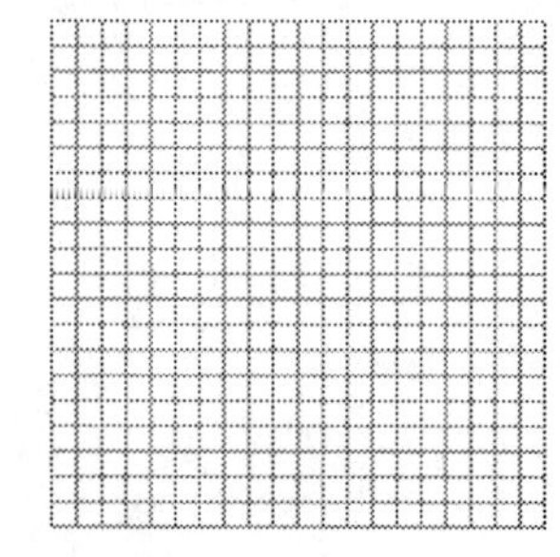

10. $f(x) = \ln(x+1) + 4$ **10.**________________

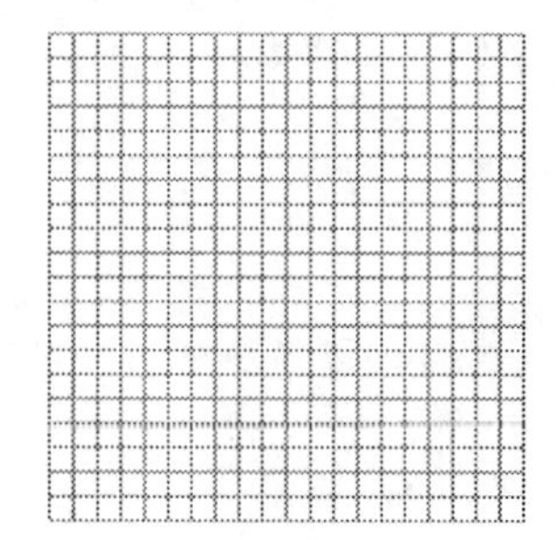

Name: Date:
Instructor: Section:

11. $f(x) = 3^{x+1} - 12$

11.________________

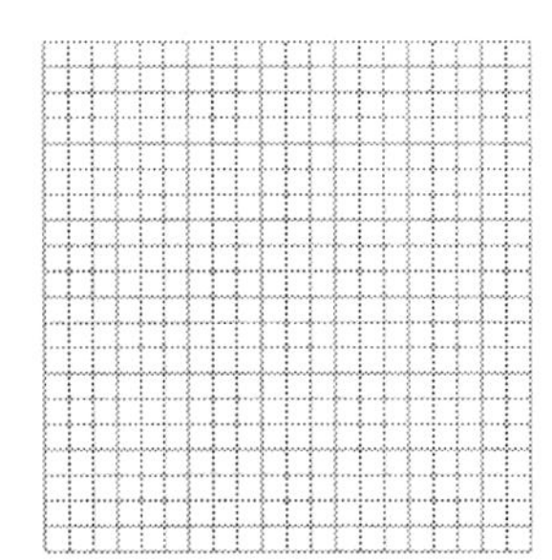

12. $f(x) = e^{x+2} - 1$

12.________________

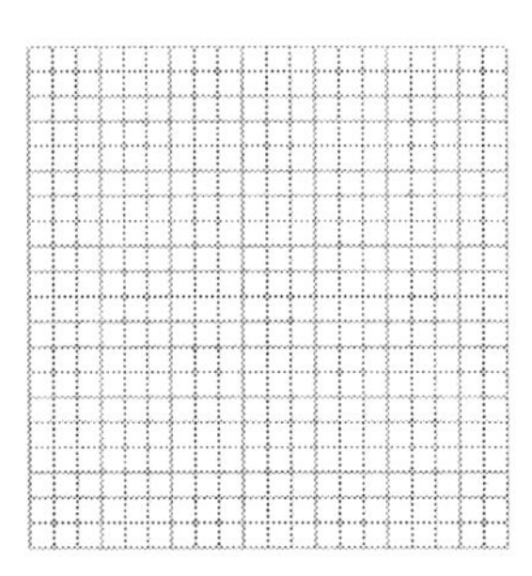

13. $f(x) = \log_2(x+8) + 3$

13.________________

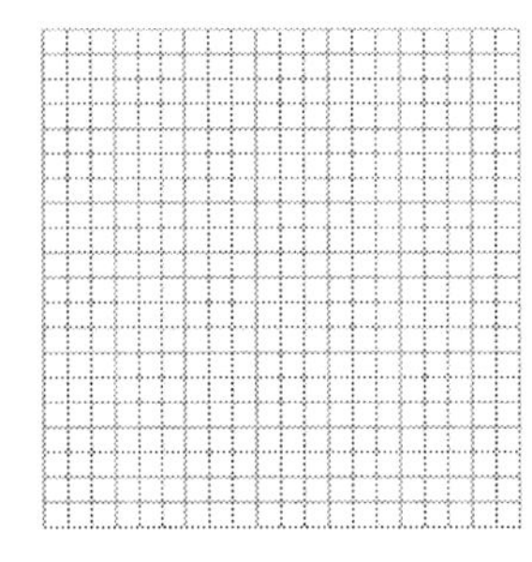

14. $f(x) = \log_6(x-4)$

14.________________

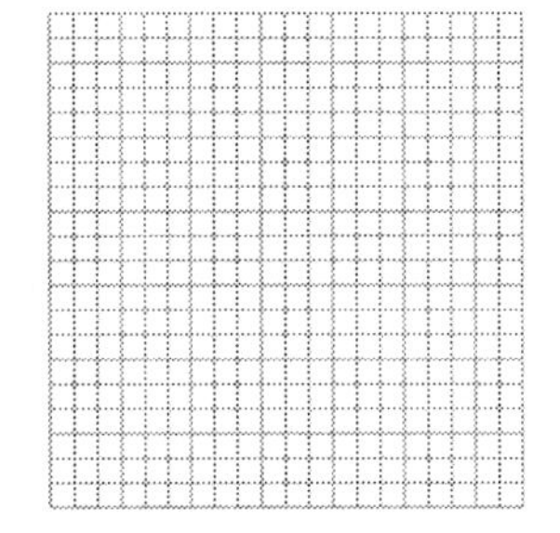

Name: Date:
Instructor: Section:

15. $f(x) = 4^{x-1}$

15.________________

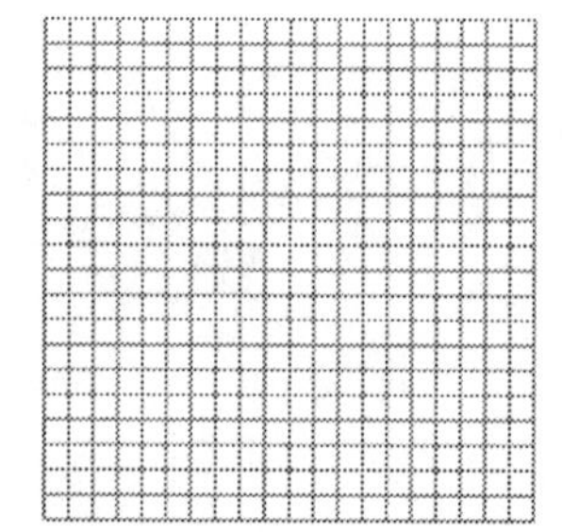

16. $f(x) = 3^{x-3} - 5$

16.________________

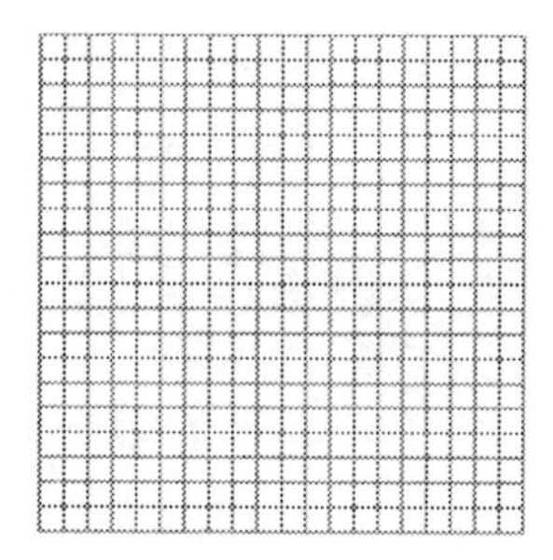

17. $f(x) = e^{x+3} - 12$

17.________________

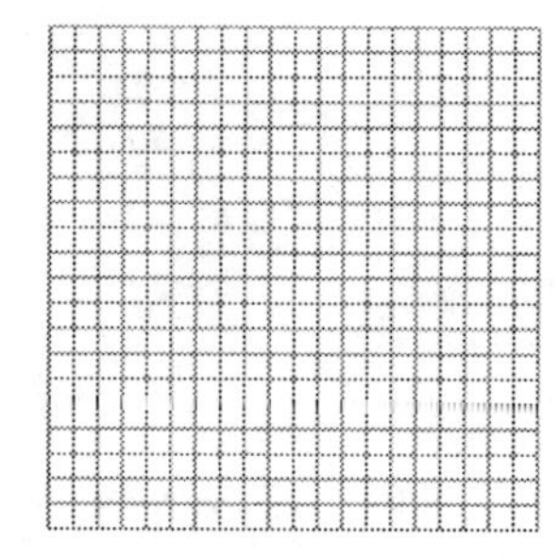

18. $f(x) = \log_4 x - 1$

18.________________

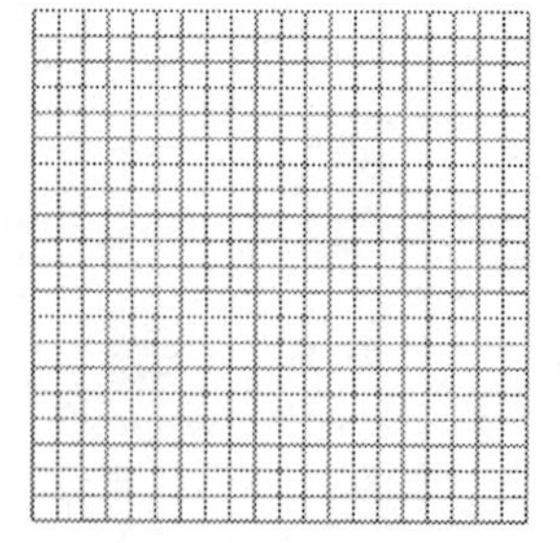

Name: Date:
Instructor: Section:

19. $f(x) = \log_3(x+4) - 3$

19.________________

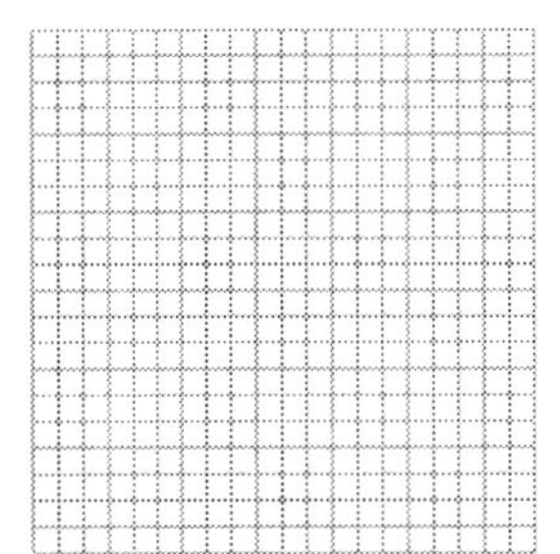

20. $f(x) = \ln(x-2) + 2$

20.

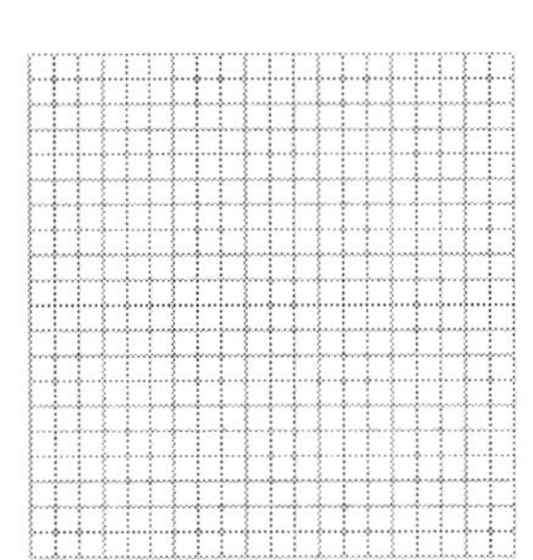

21. $f(x) = 2^{x+1} - 9$

21.

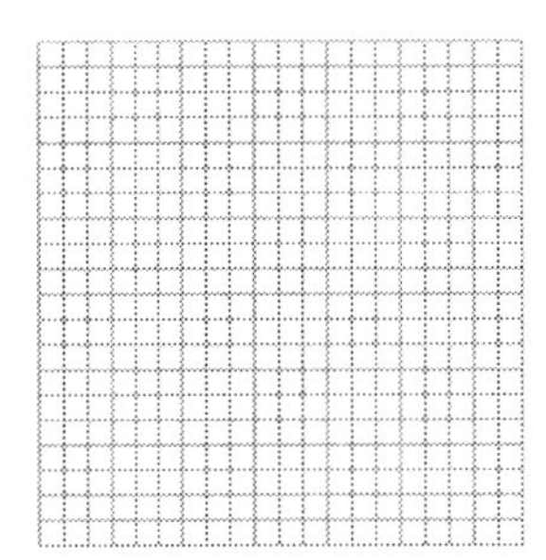

22. $f(x) = \ln(x+2) - 2$

22.

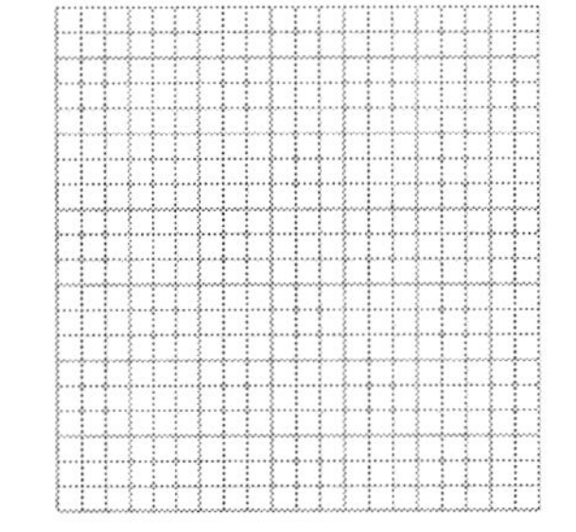

Name: Date:
Instructor: Section:

Given the graph of a function $f(x)$, graph the function $f^{-1}(x)$.

23.

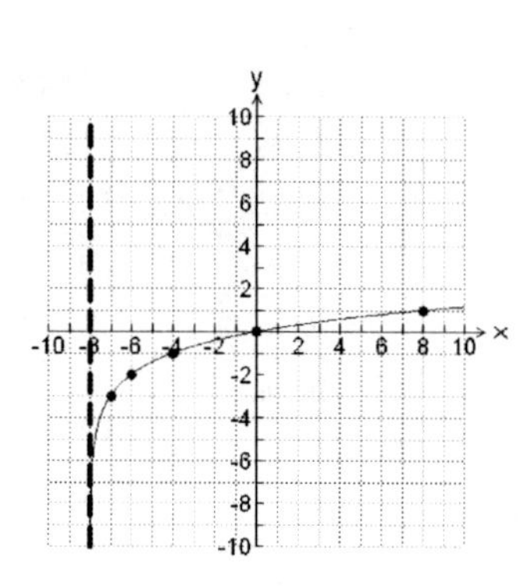

23.

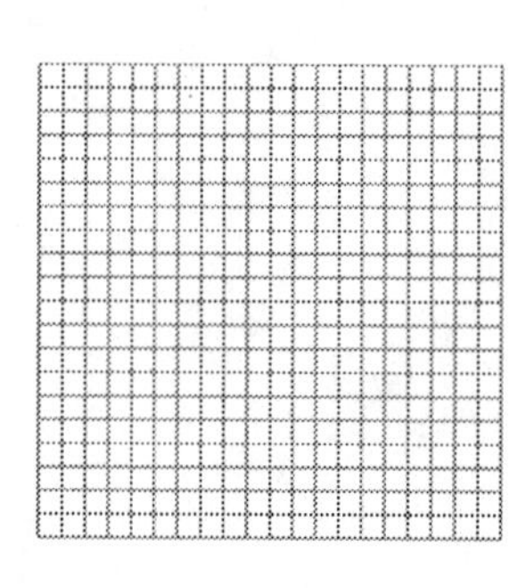

24.

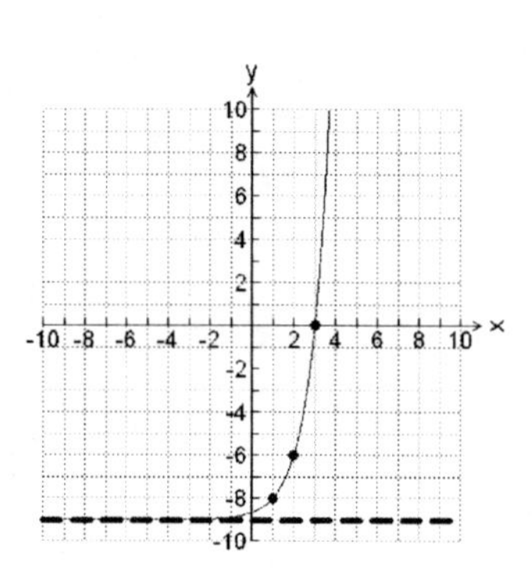

24.

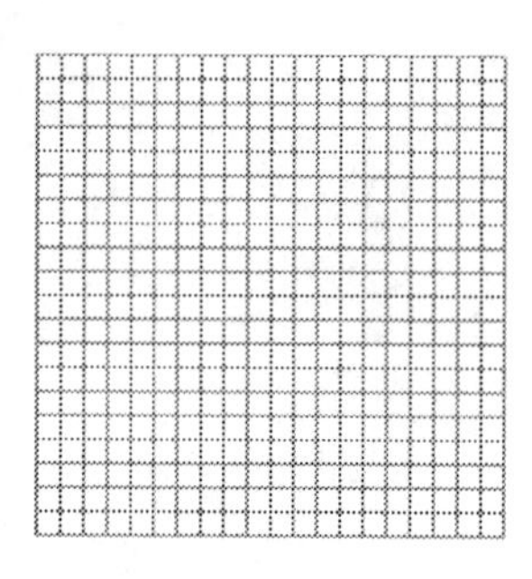

For the given function $f(x)$, find and graph $f^{-1}(x)$. Label all intercepts for $f^{-1}(x)$, as well as its asymptote.

25. $f(x) = \ln(x-6) - 2$

25.________________

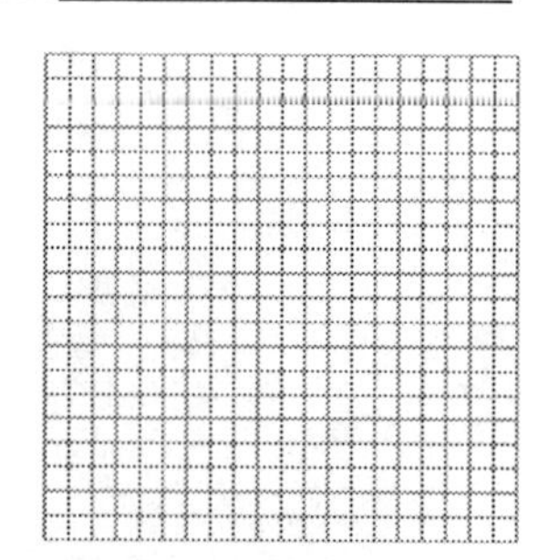

26. $f(x) = 3^{x+2}$

26.________________

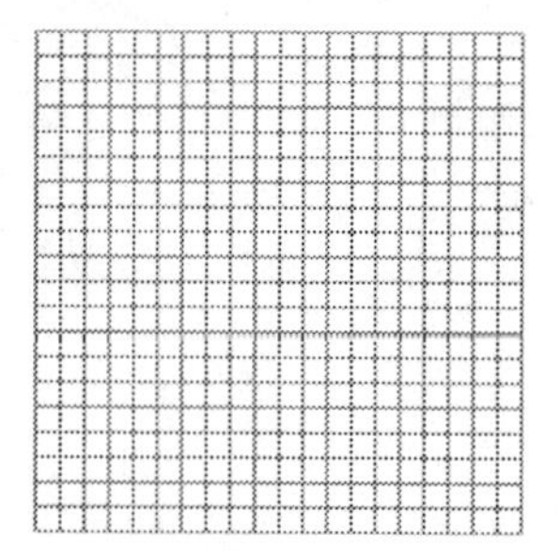

Name: Date:
Instructor: Section:

Chapter 9 CONIC SECTIONS

9.1 Parabolas

Learning Objectives

1. Graph equations of the form $y = ax^2 + bx + c$.
2. Graph equations of the form $y = a(x - h)^2 + k$.
3. Graph equations of the form $x = ay^2 + by + c$.
4. Graph equations of the form $x = a(y - k)^2 + h$.
5. Find a vertex by completing the square.
6. Find the equation of a parabola that meets the given conditions.

Key Terms

Use the most appropriate term or phrase from the given list to complete each statement in exercises 1-3.

axis of symmetry **positive** **vertex** **focus** **negative**

1. The ______________________ of a parabola of the form $y = a(x-h)^2 + k$ is at the point (h, k).

2. A parabola of the form $x = ay^2 + by + c$ opens to the left if a is ______________________.

3. A parabola of the form $y = a(x-h)^2 + k$ opens upward if a is ______________________.

Objective 1 Graph equations of the form $y = ax^2 + bx + c$.

Graph the parabola. Identify the vertex and any intercepts.

1. $y = x^2 + 11x + 18$

1.______________

Name: Date:
Instructor: Section:

2. $y = x^2 + 3x - 28$

2.________________

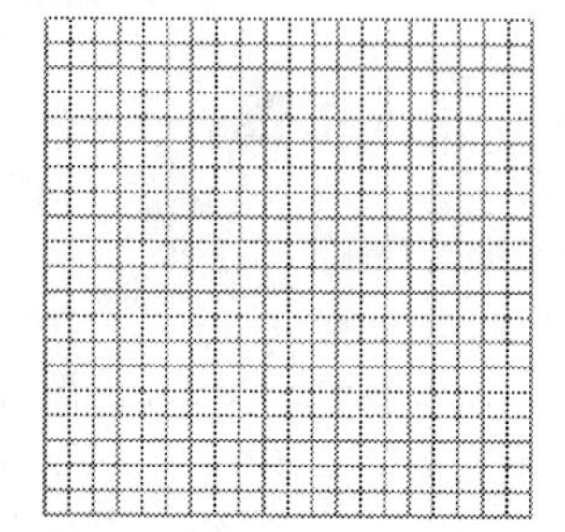

3. $y = -x^2 - 4x - 12$

3.________________

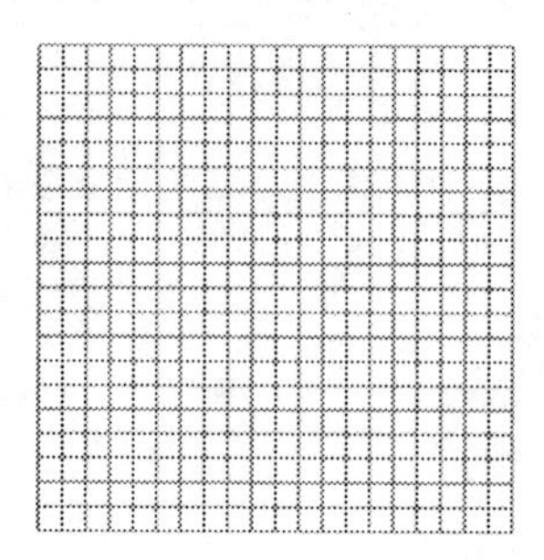

4. $y = x^2 - 4$

4.________________

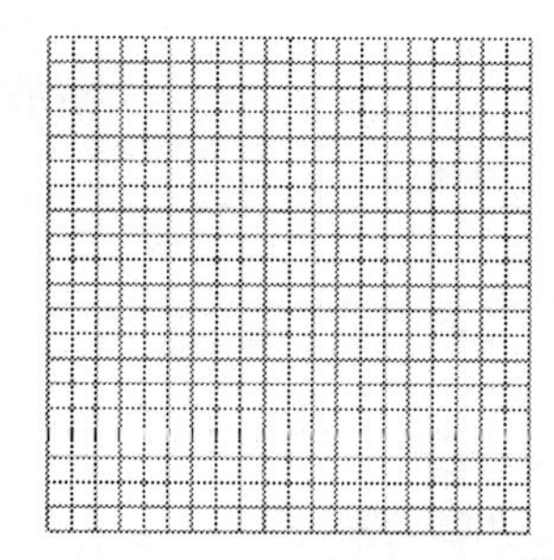

5. $y = -x^2 + 7x + 13$

5.________________

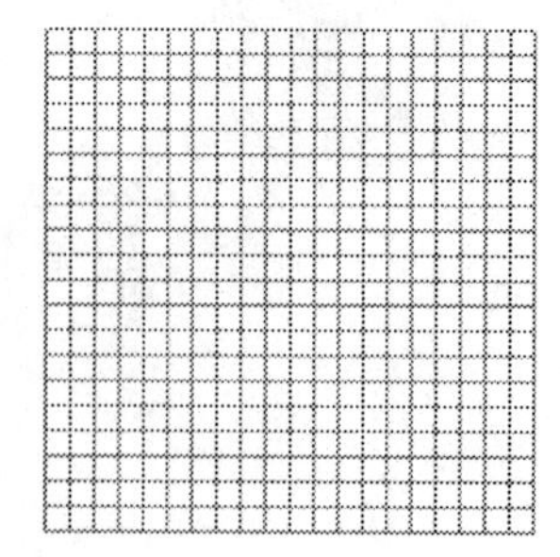

Name: Date:
Instructor: Section:

Objective 2 Graph equations of the form $y = a(x - h)^2 + k$.

Graph the parabola. Identify the vertex and any intercepts.

6. $y = (x-2)^2 - 25$

6.________________

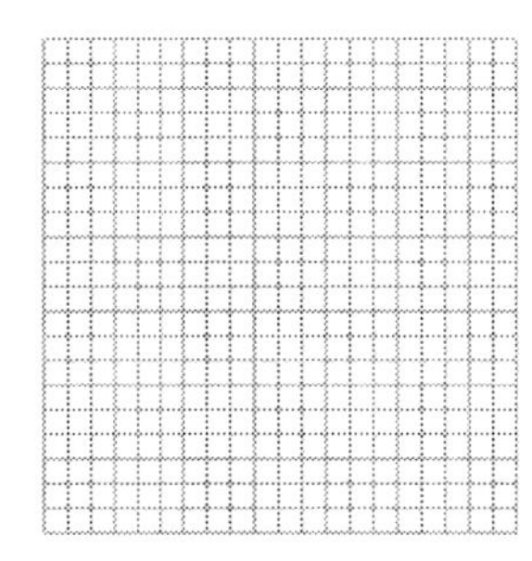

7. $y = (x-5)^2 - 9$

7.________________

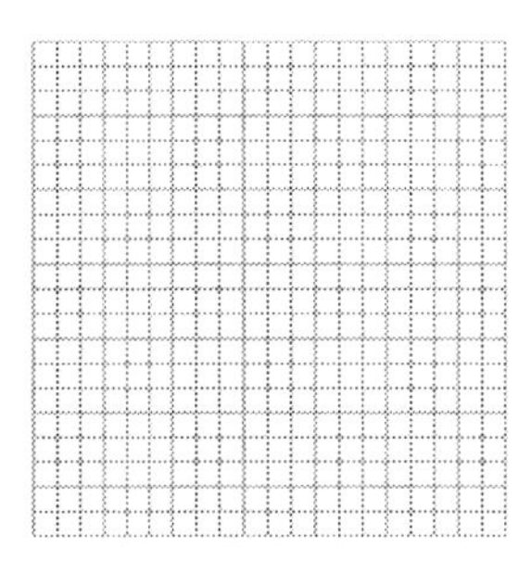

8. $y = -(x-1)^2 + 8$

8.________________

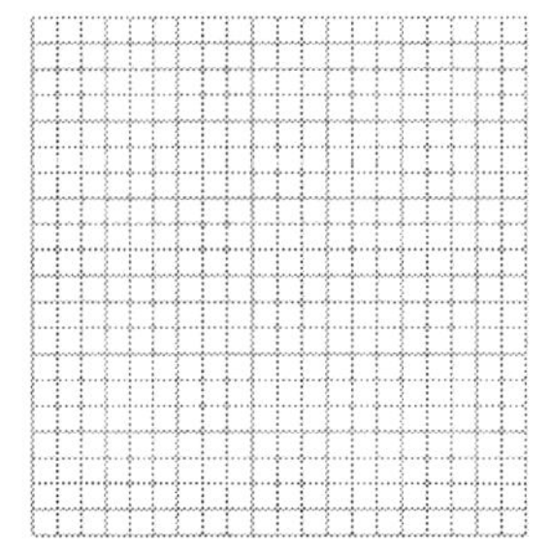

9. $y = (x+5)^2 + 3$

9.________________

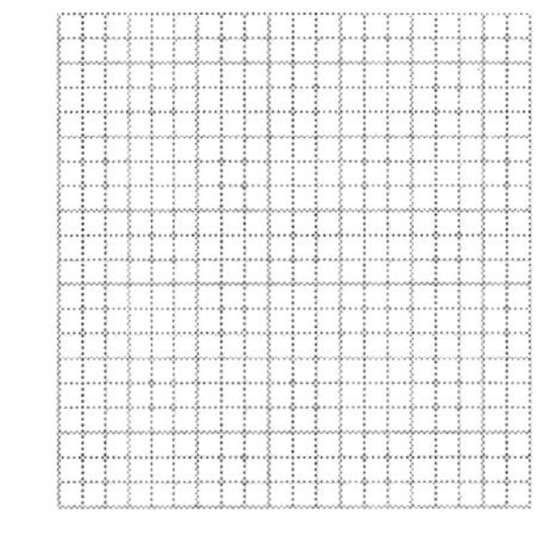

Name: Date:
Instructor: Section:

Objective 3 Graph equations of the form $x = ay^2 + by + c$.

Graph the parabola. Identify the vertex and any intercepts.

10. $x = y^2 - 5y - 6$

10.________________

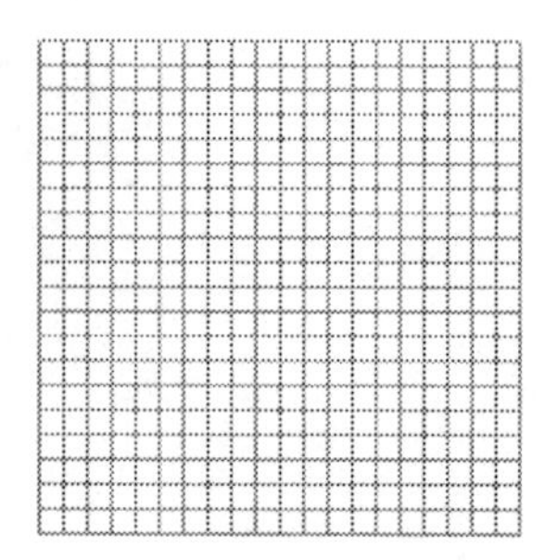

11. $x = y^2 - 9y + 14$

11.________________

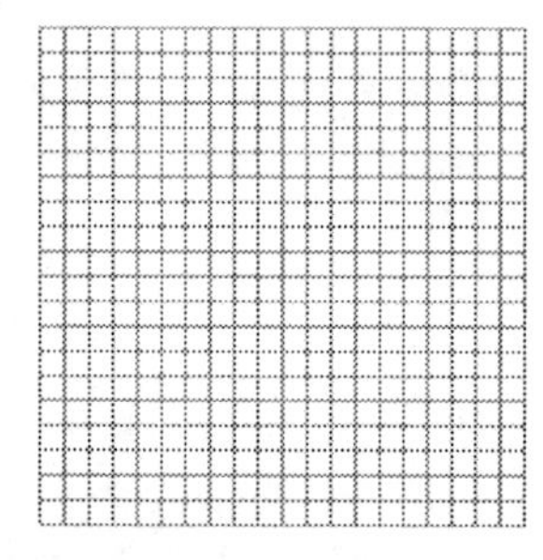

12. $x = -y^2 + 8y + 10$

12.________________

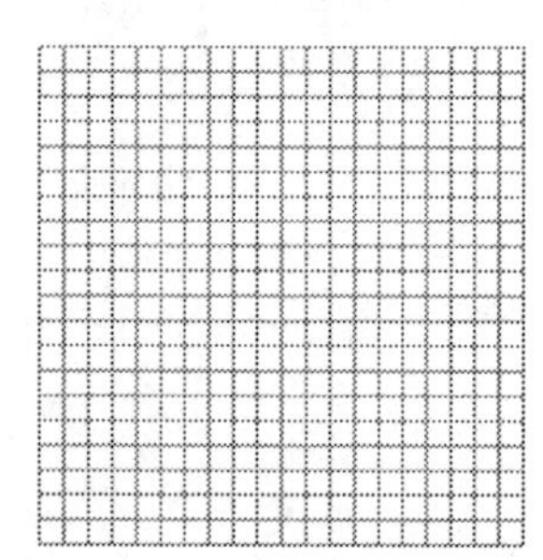

13. $x = -y^2 - 7y - 15$

13.________________

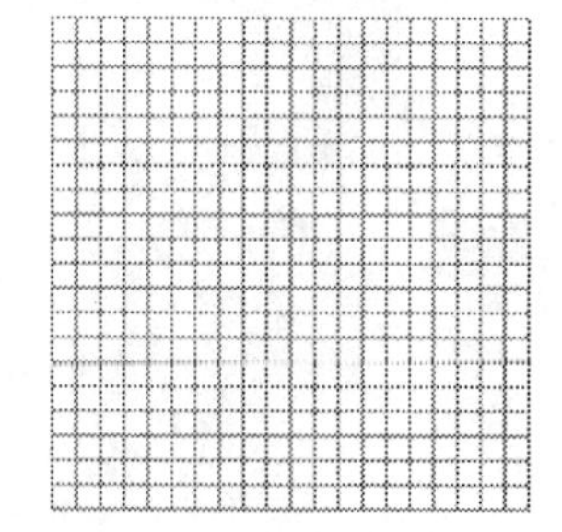

Name: Date:
Instructor: Section:

14. $x = y^2 + 7y + 9$

14.________________

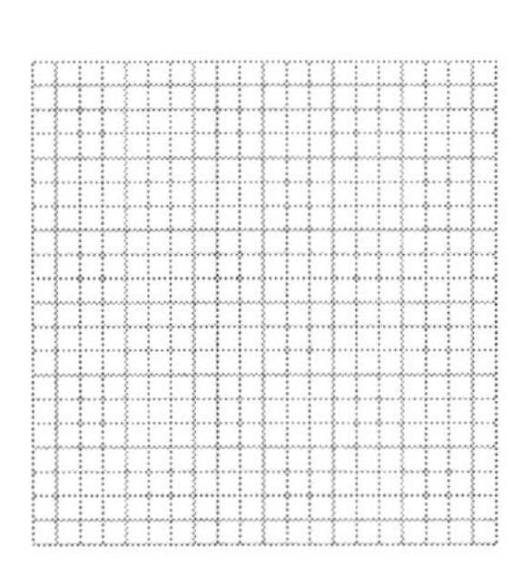

Objective 4 Graph equations of the form $x = a(y - k)^2 + h$.

Graph the parabola. Identify the vertex and any intercepts.

15. $x = (y-3)^2 - 4$

15.________________

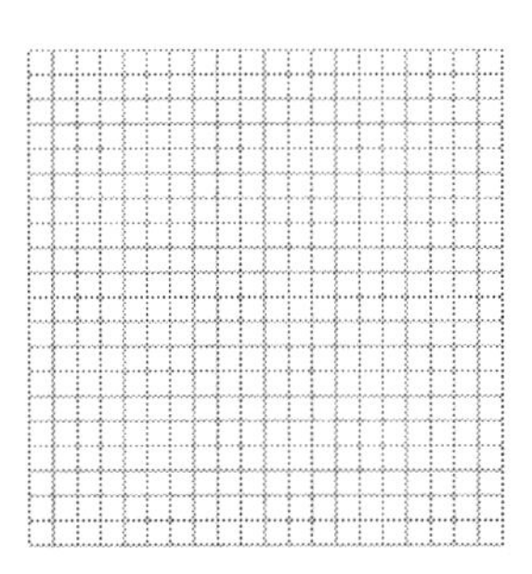

16. $x = (y-6)^2$

16.________________

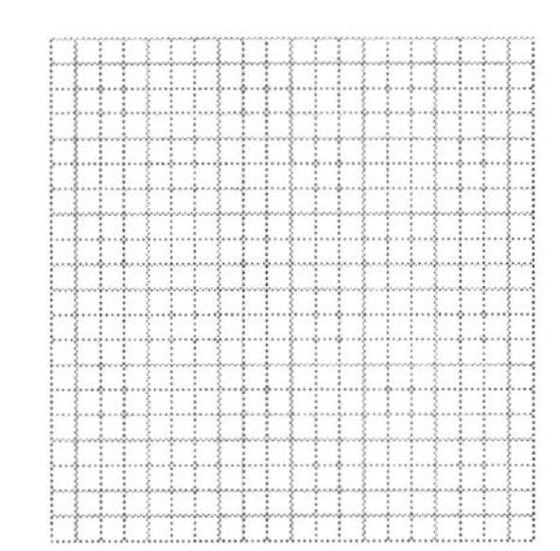

17. $x = -(y+7)^2 + 9$

17.________________

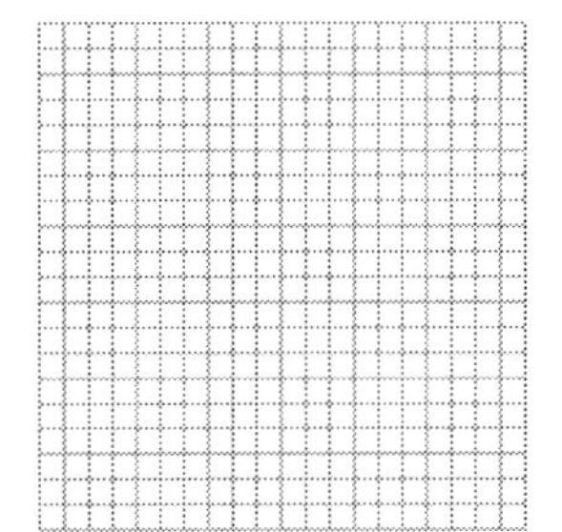

Name: Date:
Instructor: Section:

18. $x = -(y+1)^2 + 10$

18.________________

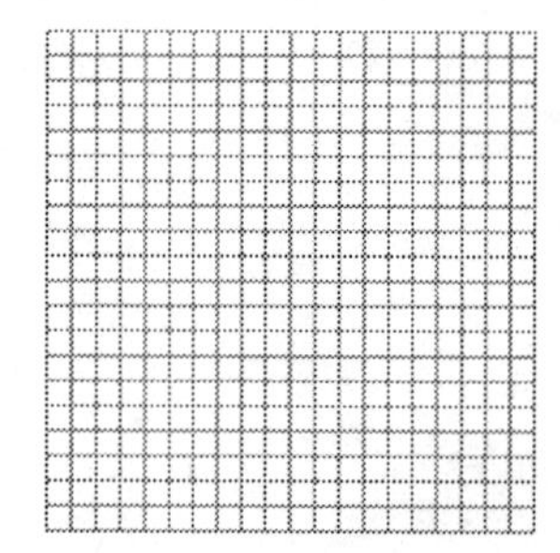

19. $x = (y+2)^2 + 6$

19.________________

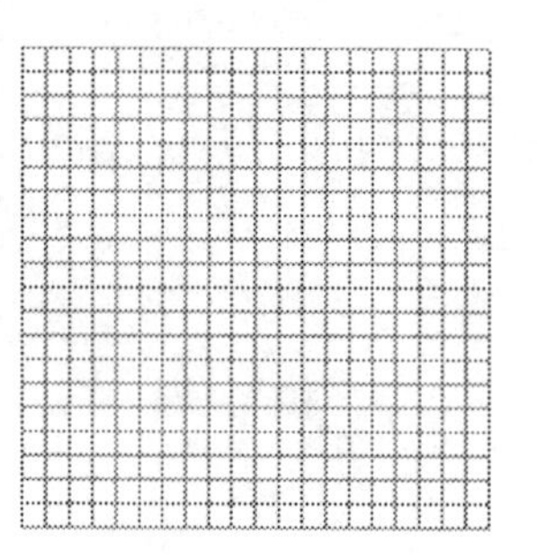

Objective 5 Find a vertex by completing the square.

Find the vertex of the parabola by completing the square.

20. $y = x^2 + 9x - 30$

20.________________

21. $y = x^2 - 5x - 45$

21.________________

Name: Date:
Instructor: Section:

22. $x = y^2 - y + 19$ **22.**________________

23. $x = y^2 + 3y - 17$ **23.**________________

Objective 6 Find the equation of a parabola that meets the given conditions.

Find the equation of the parabola that has been graphed by using the intercepts.

24. **24.**________________

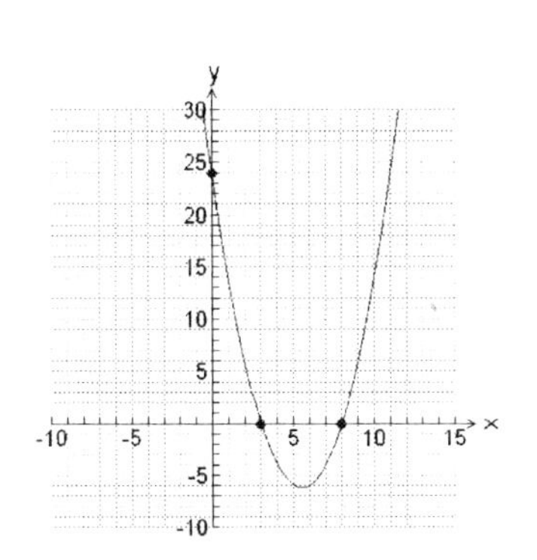

Name: Date:
Instructor: Section:

25. **25.**________________

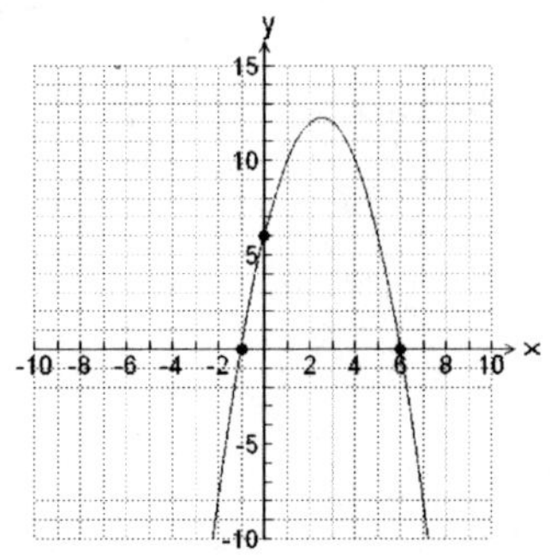

Name: Date:
Instructor: Section:

Chapter 9 CONIC SECTIONS

9.2 Circles

Learning Objectives
1 Use the distance and midpoint formulas.
2 Graph circles centered at the origin.
3 Graph circles centered at the point (h, k).
4 Find the center of a circle and its radius by completing the square.
5 Find the equation of a circle that meets the given conditions.

Key Terms
Use the most appropriate term or phrase from the given list to complete each statement in exercises 1-3.

distance **midpoint** **center** **radius** **diameter**

1. The ____________________ from the center of a circle to each point on the circle is called the radius of the circle.

2. To find the ____________________ of a line segment connecting two points, find the averages of the coordinates of the endpoints.

3. A circle is defined as the collection of all points in a plane that are a fixed distance from a point called its____________________.

Objective 1 Use the distance and midpoint formulas.

Find the distance between the given points. Round to the nearest tenth if necessary.

1. $(-11, -8)$ and $(-17, 1)$ **1.**________________

2. $(9, -14)$ and $(-31, 46)$ **2.**________________

Name: Date:
Instructor: Section:

3. $(2,\ 9)$ and $(8,\ 1)$ **3.**____________

4. $(-9,\ 4)$ and $(9,\ -5)$ **4.**____________

Find the midpoint of the line segment that connects the given points.

5. $(13,\ -13)$ and $(-7,\ -4)$ **5.**____________

6. $(-14,\ 21)$ and $(16,\ -15)$ **6.**____________

7. $(1,\ 5)$ and $(9,\ -3)$ **7.**____________

8. $(-2,\ 13)$ and $(8,\ -6)$ **8.**____________

Name: Date:
Instructor: Section:

Objective 2 Graph circles centered at the origin.

Graph the circle. State the center and radius of the circle.

9. $x^2 + y^2 = 36$

9.________________

10. $x^2 + y^2 = \frac{1}{4}$

10.________________

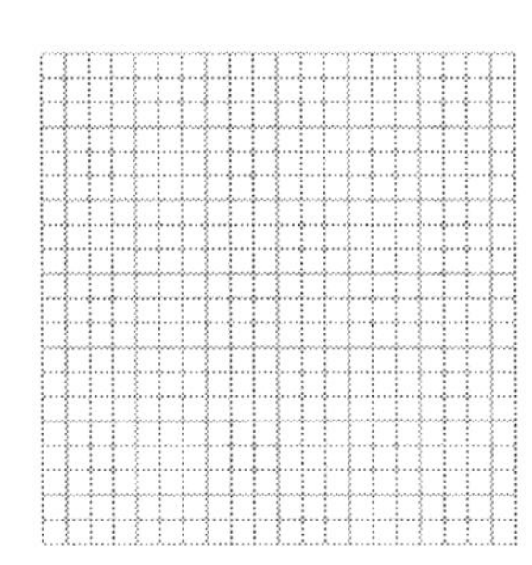

11. $x^2 + y^2 = \frac{25}{16}$

11.________________

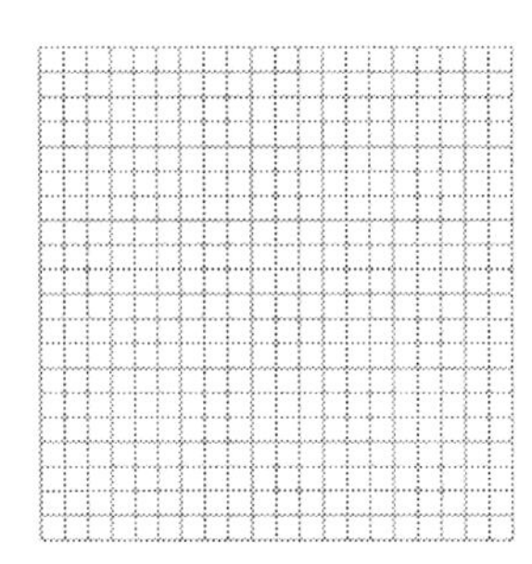

12. $x^2 + y^2 = 40$

12.________________

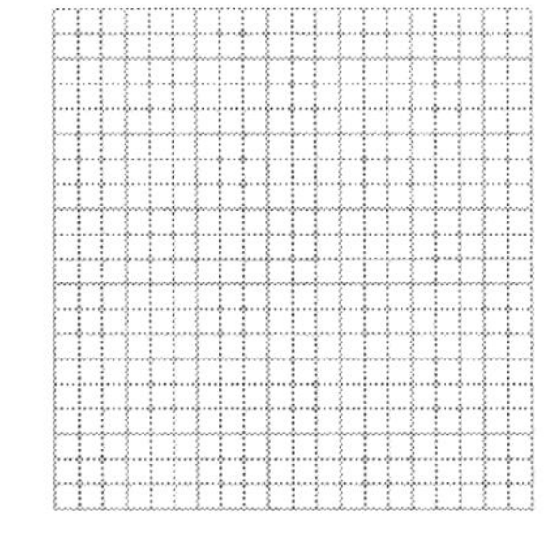

Name: Date:
Instructor: Section:

13. $x^2 + y^2 = 24$

13.________________

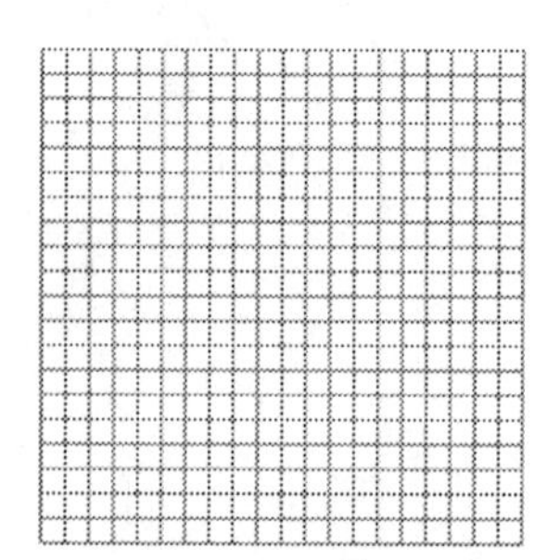

Objective 3 Graph circles centered at the point (*h*, *k*).

Graph the circle. State the center and radius of the circle.

14. $(x-6)^2 + (y+8)^2 = 100$

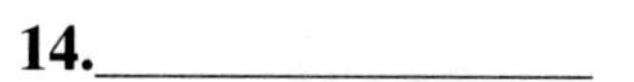

14.________________

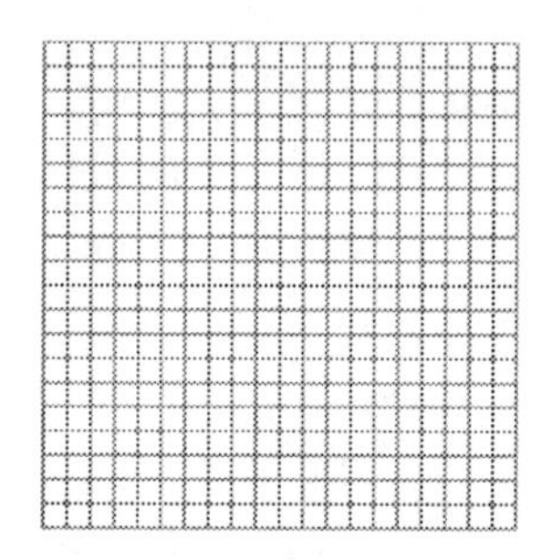

15. $(x+2)^2 + (y-7)^2 = 16$

15.________________

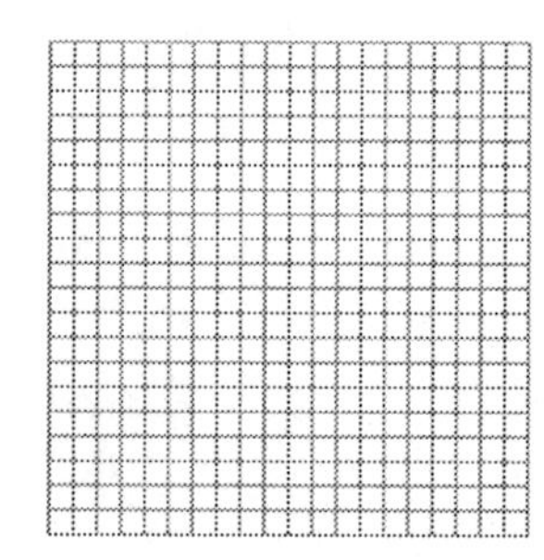

16. $\left(x+\frac{7}{2}\right)^2 + \left(y-\frac{3}{2}\right)^2 = \frac{25}{4}$

16.________________

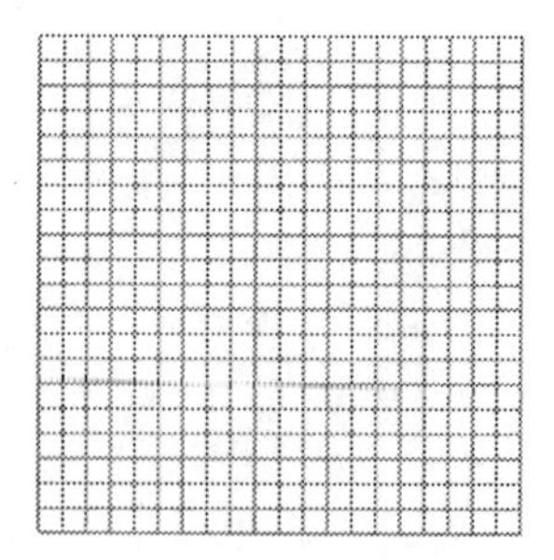

Name: Date:
Instructor: Section:

17. $\left(x-\frac{17}{2}\right)^2+\left(y+\frac{1}{2}\right)^2=\frac{9}{4}$

17.________________

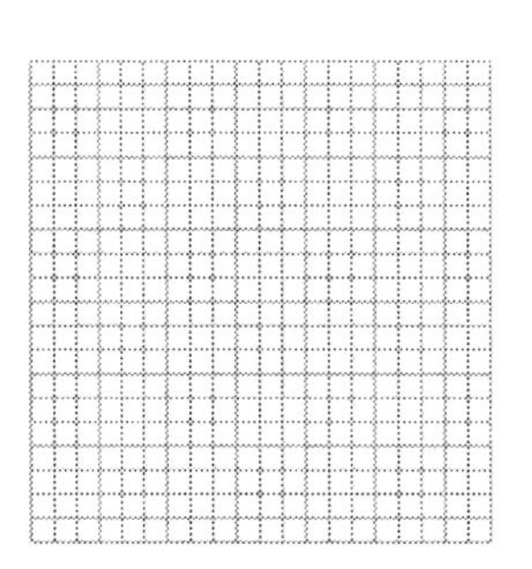

Objective 4 Find the center of a circle and its radius by completing the square.

Find the center and radius of the circle by completing the square.

18. $x^2+y^2+14x-8y-160=0$

18.________________

19. $x^2+y^2-16x-20y+20=0$

19.________________

20. $x^2+y^2+9x-13y-\frac{3}{2}=0$

20.________________

21. $x^2+y^2+5x+7y+\frac{5}{2}=0$

21.________________

Name: Date:
Instructor: Section:

22. $x^2 + y^2 - 12x - 6y + 13 = 0$ **22.**________________

Objective 5 Find the equation of a circle that meets the given conditions.

Find the equation of each circle with the given center and radius.

23. Center (0, 0), radius 7 **23.**________________

24. Center (8, 9), radius 2 **24.**________________

25. Center (3, 9), radius 5 **25.**________________

26. Center $(-7,\ 7)$, radius $\sqrt{5}$ **26.**________________

Find the equation of each circle that has a diameter with the given endpoints.

27. $(4,\ -4)$ and $(5,\ -3)$ **27.**________________

28. $(-2,\ -5)$ and $(1,\ 3)$ **28.**________________

Name: Date:
Instructor: Section:

Chapter 9 CONIC SECTIONS

9.3 Ellipses

Learning Objectives
1 Graph ellipses centered at the origin.
2 Graph ellipses centered at a point (h, k).
3 Find the center of an ellipse and the lengths of its axes by completing the square.
4 Find the equation of an ellipse that meets the given conditions.

Key Terms

Use the most appropriate term or phrase from the given list to complete each statement in exercises 1-3.

vertices **focus** **major** **ellipse** **minor** **co-vertices**

1. The center of a(n) ________________ is the point located midway between the two foci.

2. The ________________ axis is the shorter of an ellipse's two axes.

3. The endpoints of the minor axis of an ellipse are called the ________________.

Objective 1 Graph ellipses centered at the origin.

Graph the ellipse. Give the coordinates of the center, as well as the values of a and b.

1. $\frac{x^2}{16}+\frac{y^2}{81}=1$

1.________________

Name: Date:
Instructor: Section:

2. $\dfrac{x^2}{100}+\dfrac{y^2}{49}=1$

2.________________

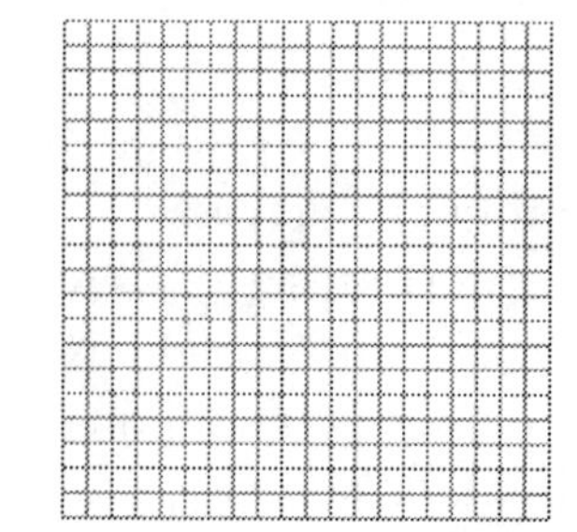

3. $\dfrac{x^2}{18}+\dfrac{y^2}{49}=1$

3.________________

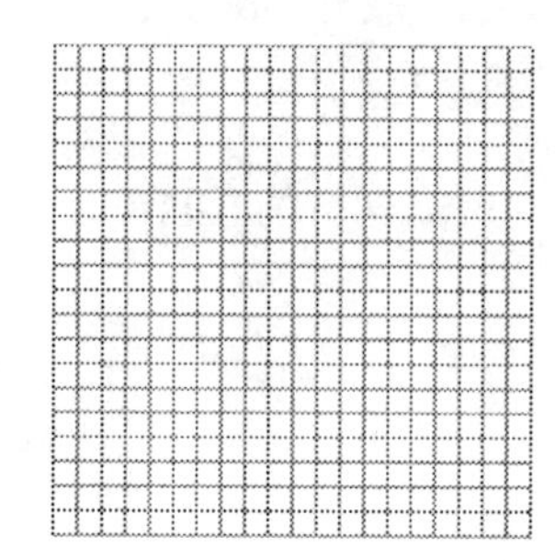

4. $\dfrac{x^2}{20}+\dfrac{y^2}{9}=1$

4.________________

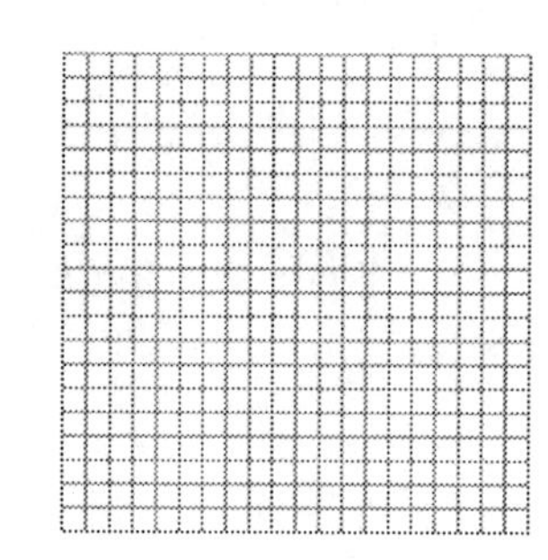

5. $8x^2+10y^2=200$

5.________________

Name: Date:
Instructor: Section:

6. $6x^2 + 24y^2 = 600$

6.________________

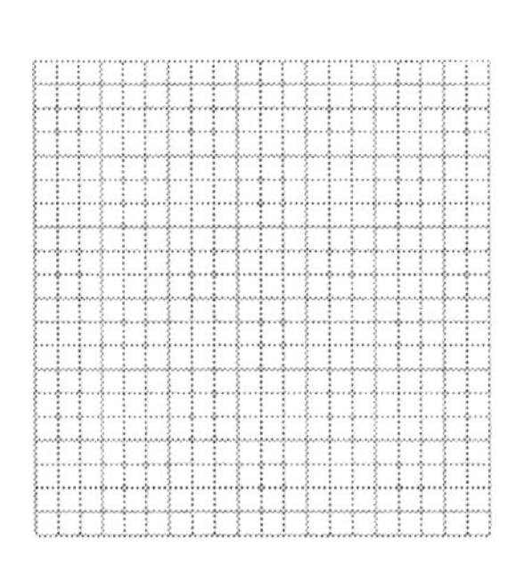

Objective 2 Graph ellipses centered at a point (h, k).

Graph the ellipse. Give the coordinates of the center, as well as the values of a and b.

7. $\frac{(x-3)^2}{64} + \frac{(y+2)^2}{16} = 1$

7.________________

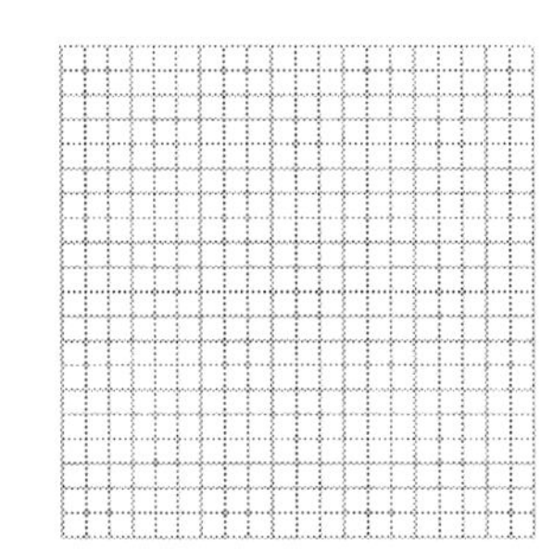

8. $\frac{(x+4)^2}{81} + \frac{(y-5)^2}{36} = 1$

8.________________

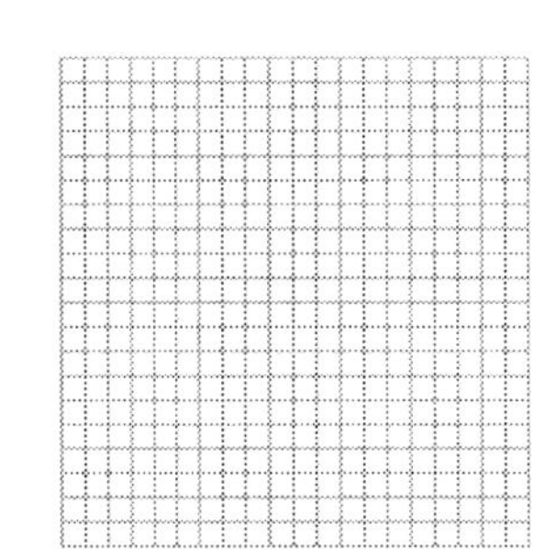

9. $\frac{(x+4)^2}{25} + \frac{(y+9)^2}{100} = 1$

9.________________

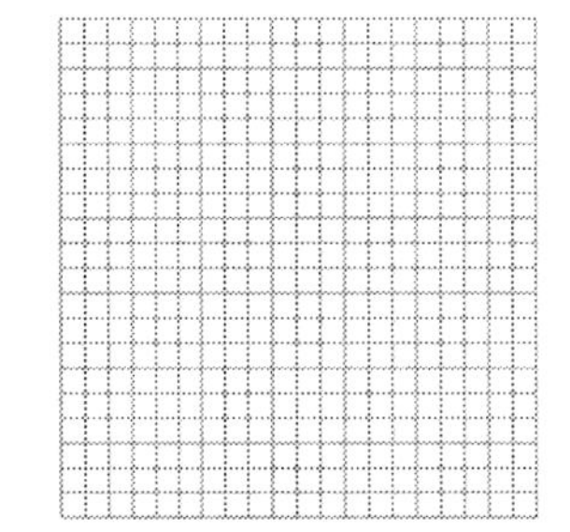

Name: Date:
Instructor: Section:

10. $\frac{(x+3)^2}{100}+\frac{(y+2)^2}{81}=1$

10.________________

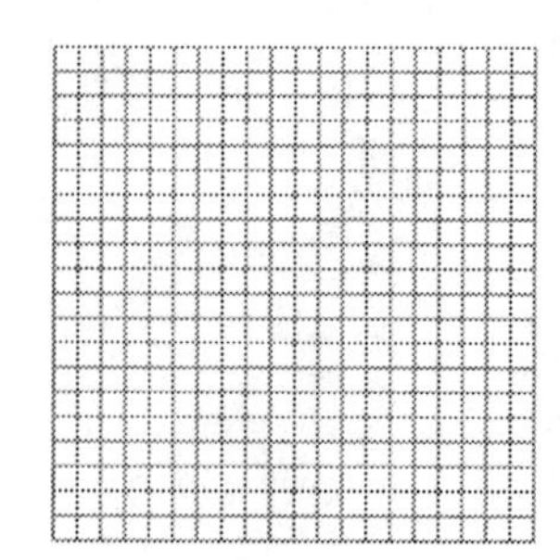

11. $\frac{(x+3)^2}{9}+\frac{(y-4)^2}{32}=1$

11.________________

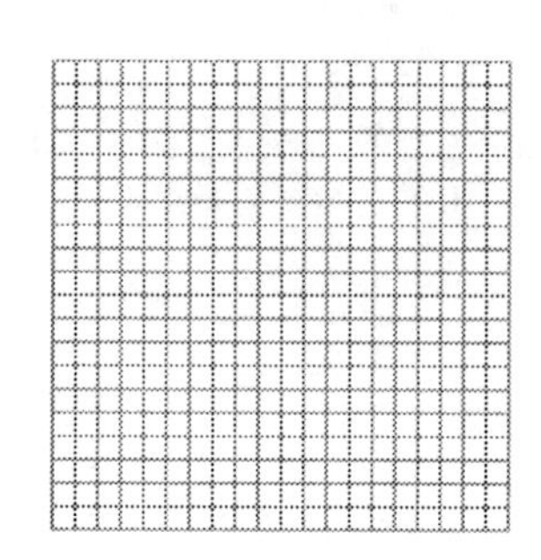

12. $\frac{(x-5)^2}{49}+\frac{(y+3)^2}{40}=1$

12.________________

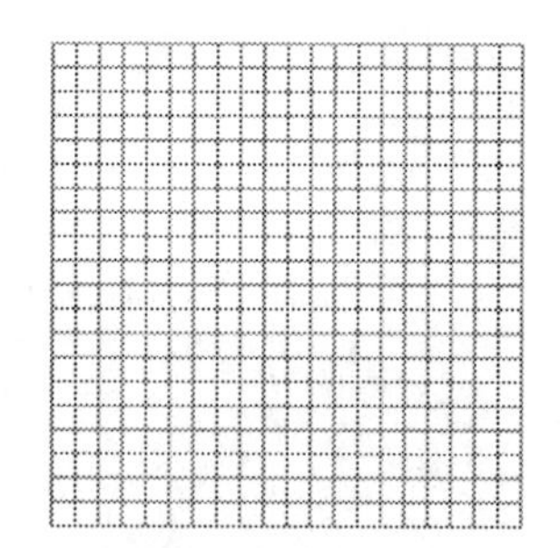

Name: Date:
Instructor: Section:

Objective 3 Find the center of an ellipse and the lengths of its axes by completing the square.

Graph the ellipse. Give the coordinates of the center, as well as the values of a and b.

13. $36x^2 + 100y^2 - 360x + 600y - 1800 = 0$

13.________________

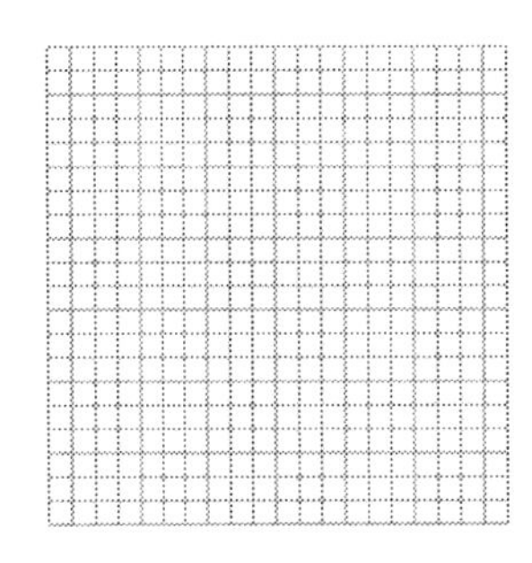

14. $48x^2 + 27y^2 + 768x - 432y + 4368 = 0$

14.________________

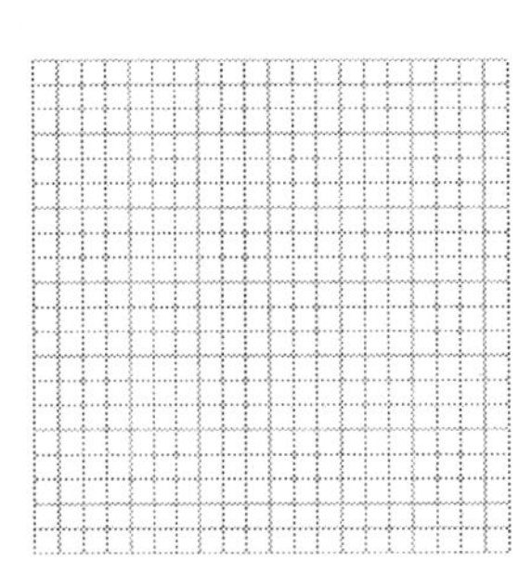

15. $25x^2 + y^2 - 150x + 14y + 174 = 0$

15.________________

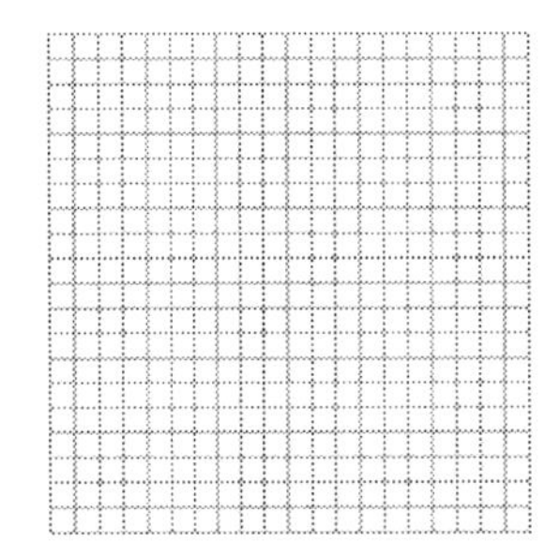

Name: Date:
Instructor: Section:

Objective 4 Find the equation of an ellipse that meets the given conditions.

Find the standard-form equation of the ellipse that meets the given conditions.

16. Center: $(6, -4)$ **16.**________________

Major axis: horizontal
Length of major axis: 18
Length of minor axis: 6

17. Center: $(7, 1)$ **17.**________________

Major axis: vertical
Length of major axis: 12
Length of minor axis: 8

18. Center: $(-4, 5)$ **18.**________________

Major axis: horizontal
Length of major axis: 16
Length of minor axis: 6

19. **19.**________________

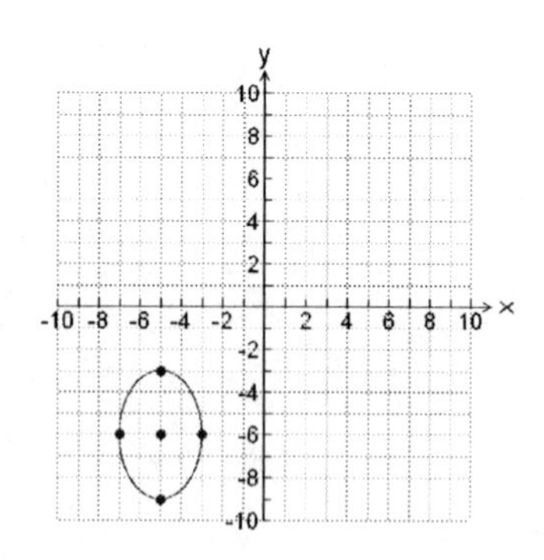

Name: Date:
Instructor: Section:

20.

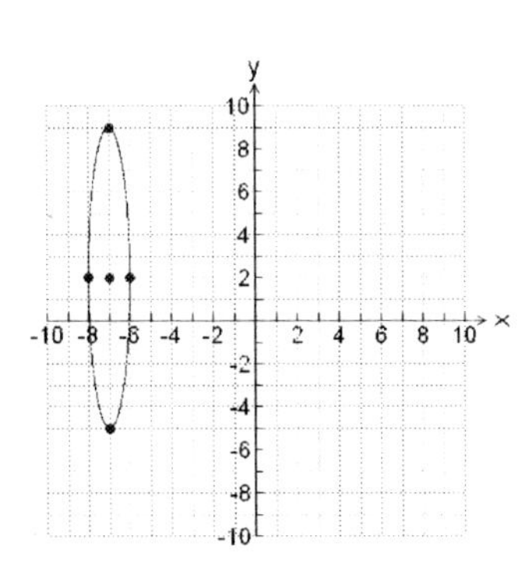

20.________________

21.

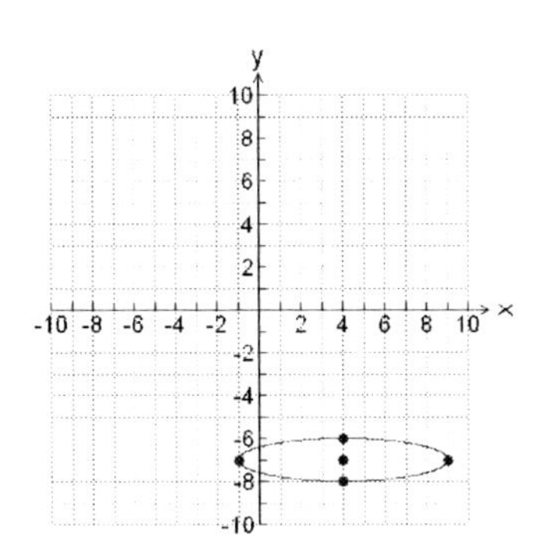

21.________________

22.

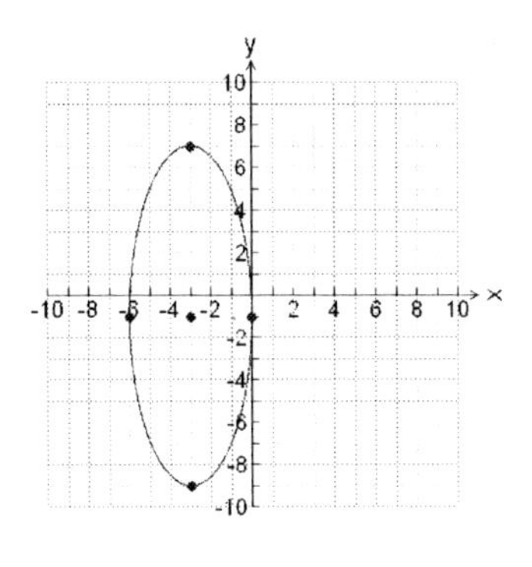

22.________________

23.

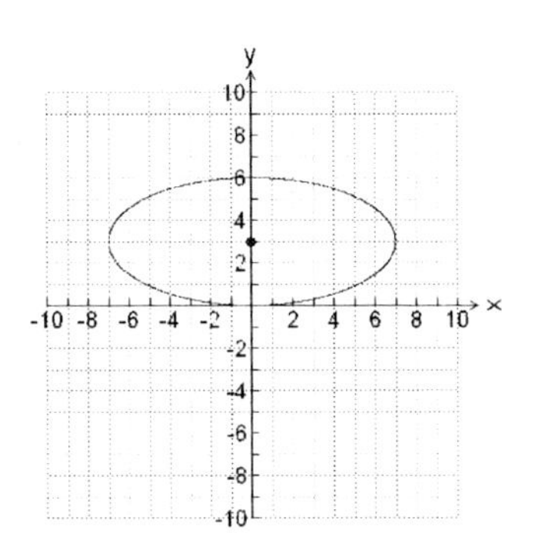

23.________________

Name: Date:
Instructor: Section:

24.

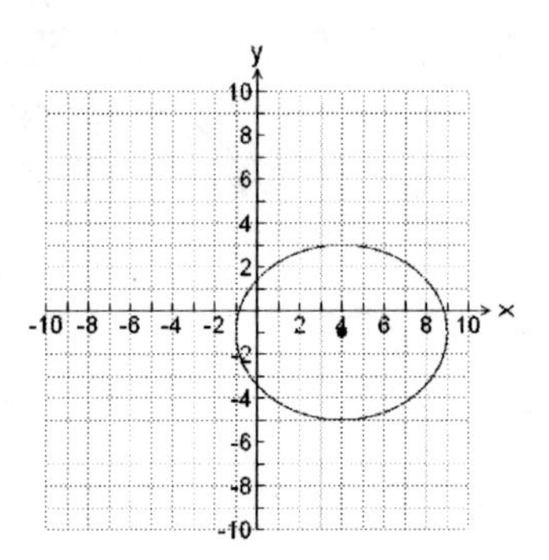

24.________________

Name: Date:
Instructor: Section:

Chapter 9 CONIC SECTIONS

9.4 Hyperbolas

Learning Objectives

1. Graph hyperbolas centered at the origin.
2. Graph hyperbolas centered at a point (h, k).
3. Find the center of a hyperbola and the lengths of its axes by completing the square.
4. Find the equation of a hyperbola that meets the given conditions.

Key Terms

Use the most appropriate term from the given list to complete each statement in exercises 1-4.

hyperbola	**branches**	**center**	**vertices**
transverse	**asymptote**	**conjugate**	**foci**

1. The ________________ axis is the line segment between the vertices of a hyperbola.

2. An ________________ of a hyperbola shows how the hyperbola behaves away from the center of the hyperbola.

3. The center of a hyperbola is the point located midway between the ________________.

4. The ________________ axis of a hyperbola is perpendicular to the transverse axis.

Objective 1 Graph hyperbolas centered at the origin.

Graph the hyperbola. Give the coordinates of the center, as well as the values of a and b.

1. $\frac{y^2}{25}-\frac{x^2}{81}=1$

1.________________

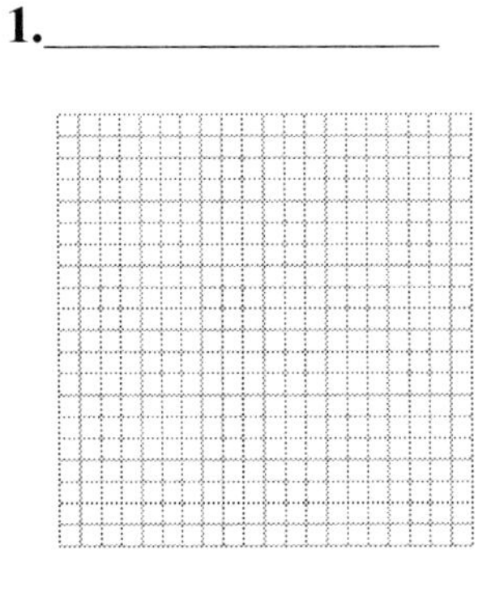

Name: Date:
Instructor: Section:

2. $\frac{x^2}{49} - \frac{y^2}{36} = 1$

2.________________

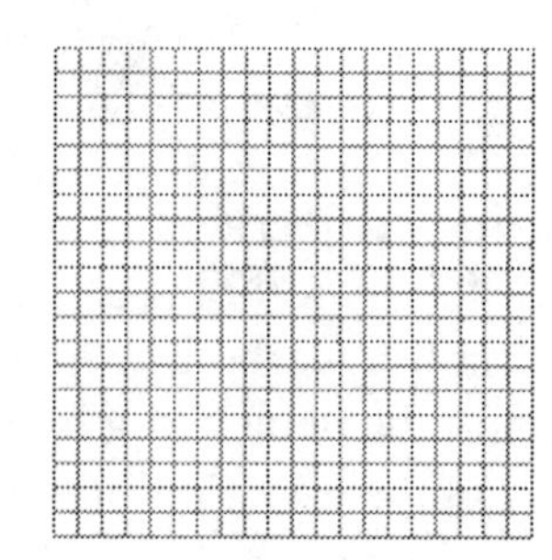

3. $8y^2 - 25x^2 = 200$

3.________________

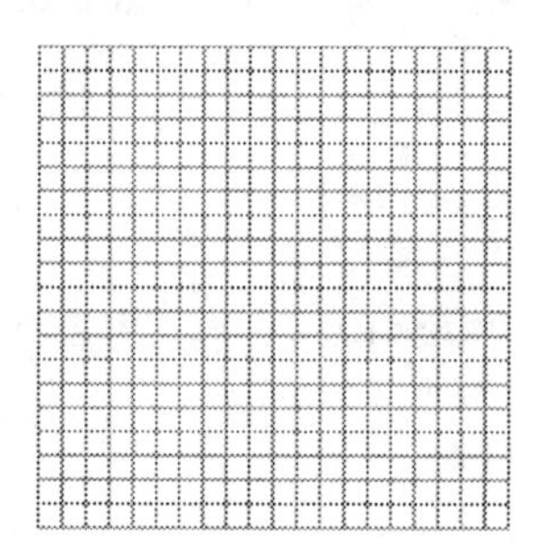

4. $7x^2 - 16y^2 = 112$

4.________________

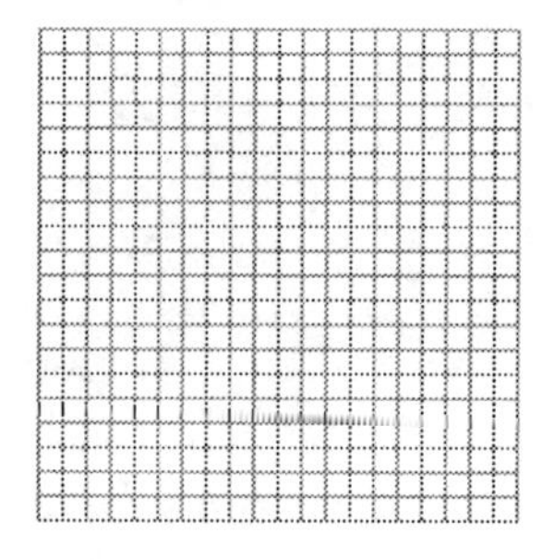

Objective 2 Graph hyperbolas centered at a point (h, k).

Graph the hyperbola. Give the coordinates of the center, as well as the values of a and b.

5. $\frac{(x+3)^2}{4} - \frac{(y-4)^2}{9} = 1$

5.________________

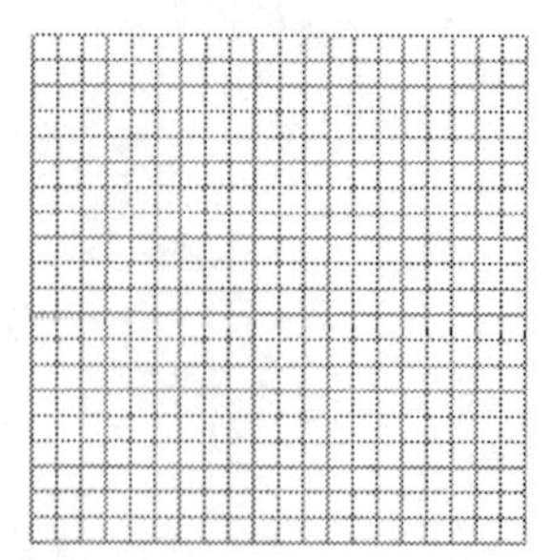

Name: Date:
Instructor: Section:

6. $\dfrac{y^2}{25}-\dfrac{(x-5)^2}{9}=1$

6.________________

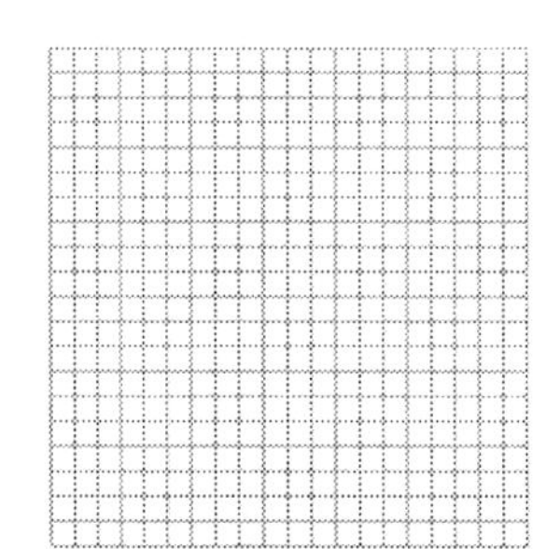

7. $\dfrac{(y-3)^2}{16}-\dfrac{(x+6)^2}{25}=1$

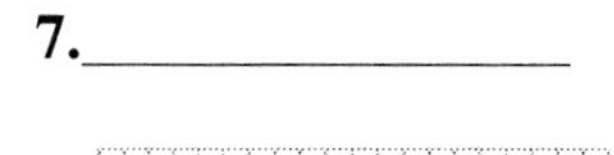

7.________________

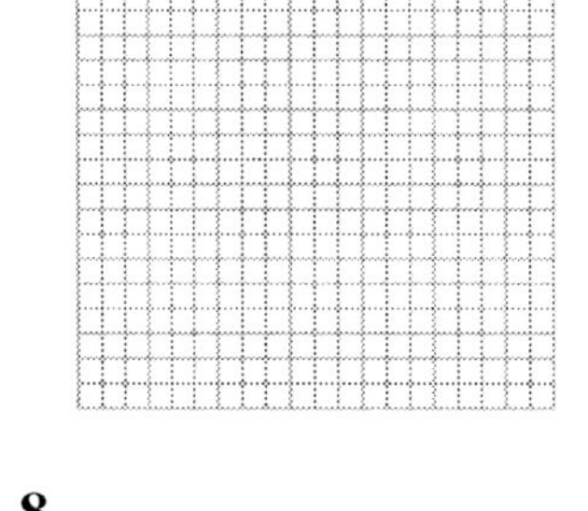

8. $\dfrac{(x-4)^2}{9}-\dfrac{(y+1)^2}{64}=1$

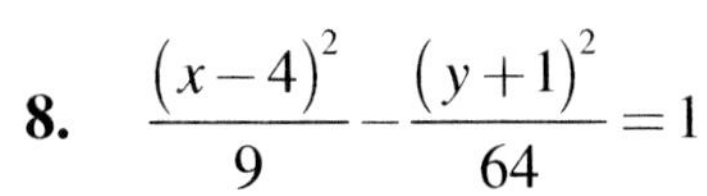

8.________________

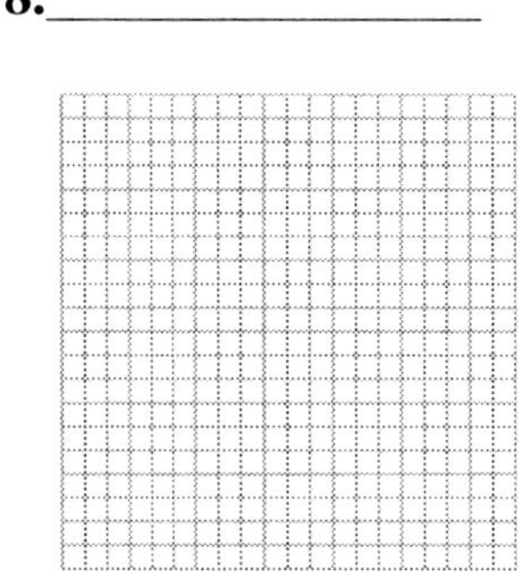

9. $\dfrac{(x-8)^2}{3}-\dfrac{(y-6)^2}{8}=1$

9.________________

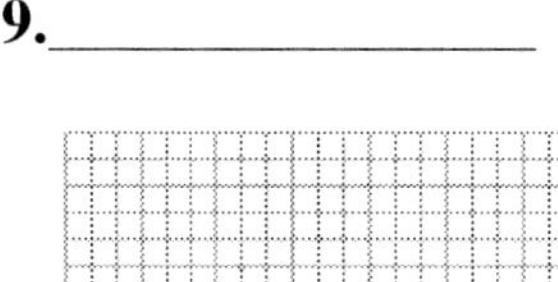

Name: Date:
Instructor: Section:

10. $\dfrac{(y-3)^2}{10}-\dfrac{(x+4)^2}{15}=1$

10.________________

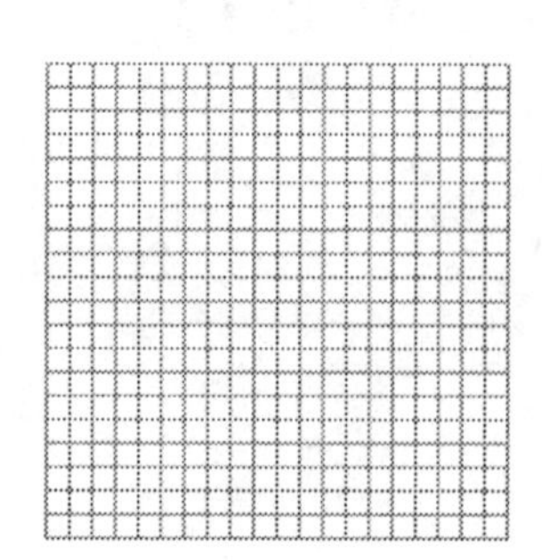

Objective 3 Find the center of a hyperbola and the lengths of its axes by completing the square.

Graph the hyperbola. Give the coordinates of the center, as well as the values of a and b.

11. $36x^2-25y^2+432x+396=0$

11.________________

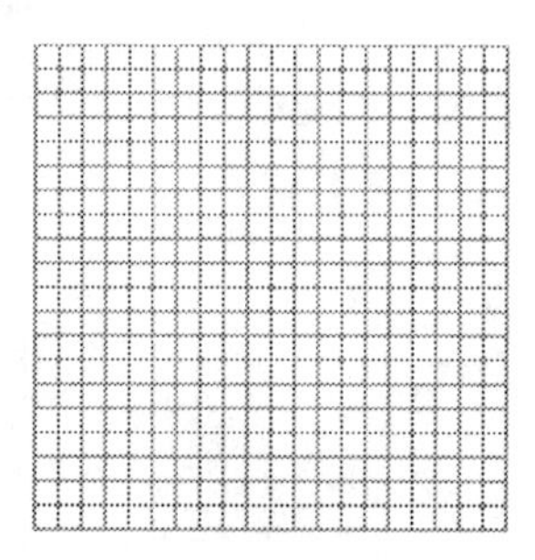

12. $-100x^2+4y^2+800x-2000=0$

12.________________

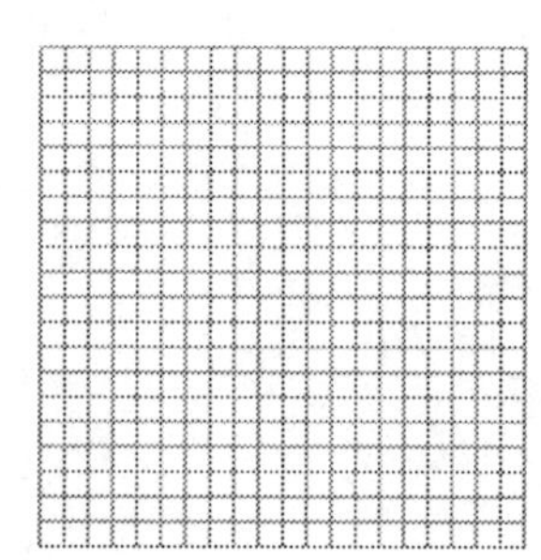

13. $9x^2-16y^2-72x-32y-16=0$

13.________________

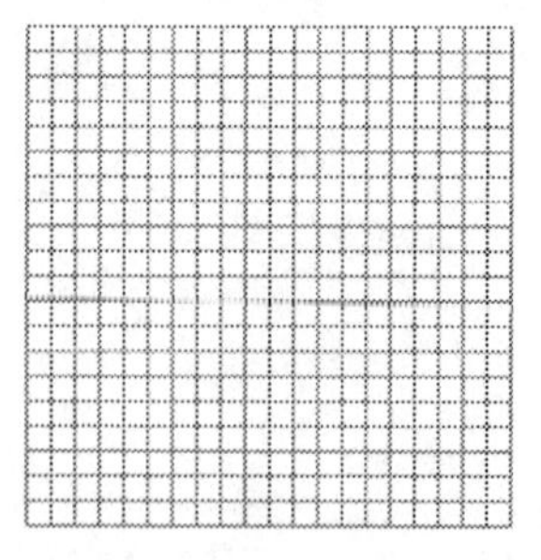

Name: Date:
Instructor: Section:

Objective 4 Find the equation of a hyperbola that meets the given conditions.

Find the standard-form equation of the hyperbola that meets the given conditions.

14. Center: $(-3, \ -1)$ **14.**________________

Transverse axis: horizontal
Length of transverse axis: 12
Length of conjugate axis: 8

15. Center: $(8, \ -9)$ **15.**________________

Transverse axis: horizontal
Length of transverse axis: 16
Length of conjugate axis: 18

Find the standard-form equation of the hyperbola whose graph is shown.

16. **16.**

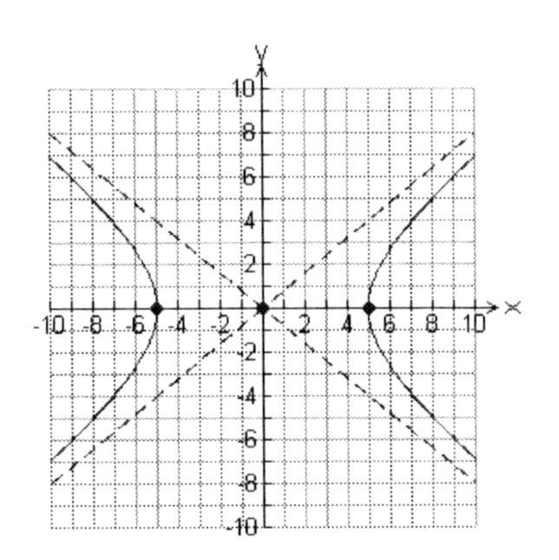

17. **17.**________________

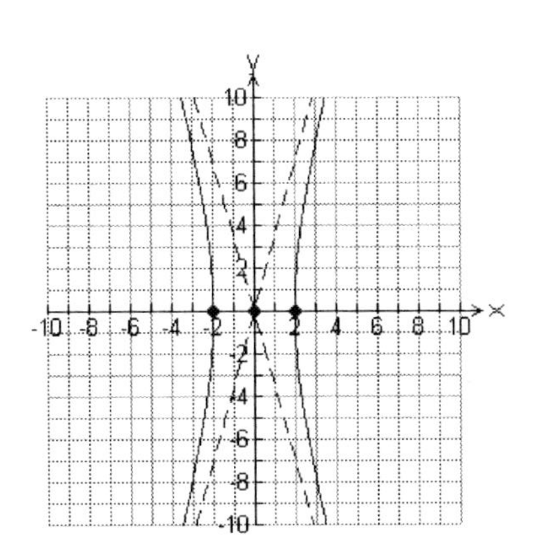

Name: Date:
Instructor: Section:

18. **18.**________________

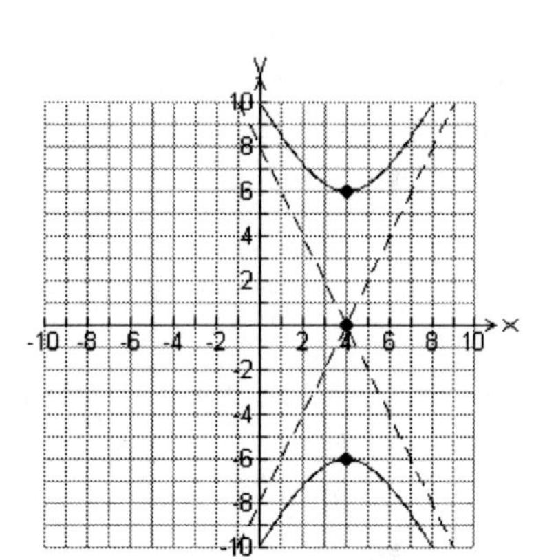

19. **19.**________________

20. **20.**________________

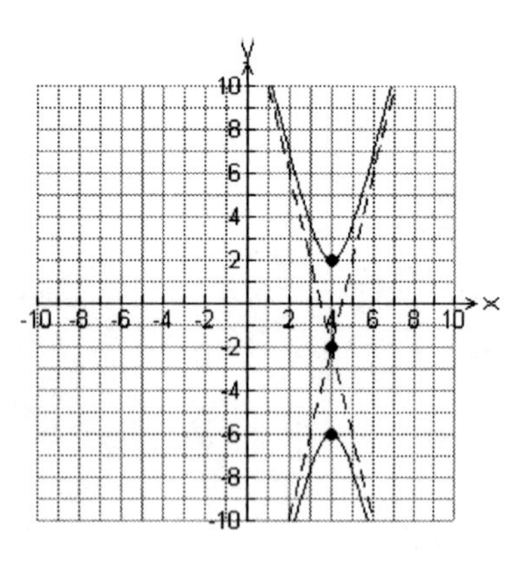

Name: Date:
Instructor: Section:

Chapter 9 CONIC SECTIONS

9.5 Nonlinear Systems of Equations

Learning Objectives
1. Solve nonlinear systems by using the substitution method.
2. Solve nonlinear systems by using the addition method.
3. Solve applications of nonlinear systems.

Objective 1 Solve nonlinear systems by using the substitution method.

Solve by the substitution method.

1. $x^2 + y^2 = 65$
$-x + 3y = 5$

1.________________

2. $x^2 + (y-4)^2 = 225$
$-x + 7y = 103$

2.________________

3. $x^2 + 4y^2 = 25$
$3x - 8y = 25$

3.________________

4. $5x^2 + 2y^2 = 23$
$3x + y = 0$

4.________________

Name: Date:
Instructor: Section:

5. $x = y^2 + 7y - 20$
$x - 9y = -17$

5.________________

6. $x = y^2 - 4y - 18$
$x - 7y = -36$

6.________________

7. $y^2 - x^2 = 7$
$y = x + 7$

7.________________

8. $y^2 - 4x^2 = 9$
$y = x + 3$

8.________________

9. $3x^2 - 5y^2 = 30$
$x - y = 8$

9.________________

10. $4y^2 - 9x^2 = 27$
$x - y = -2$

10.________________

Name: Date:
Instructor: Section:

11. $x^2 + y^2 = 100$
$y = -\frac{1}{8}x^2 + 2$

11.________________

12. $49x^2 + 4y^2 = 193$
$y = x^2 + 5$

12.________________

13. $9x^2 + 16y^2 = 144$
$x = y^2 - 4$

13.________________

14. $y = 2x^2 + 5x - 5$
$y = x^2 + 4x + 7$

14.________________

15. $y = 5x^2 - 23x + 19$
$y = 4x^2 - 7x - 41$

15.________________

Name: Date:
Instructor: Section:

Objective 2 Solve nonlinear systems by using the addition method.

Solve by the addition method.

16. $x^2 + y^2 = 37$
$x^2 - y^2 = 35$

16.________________

17. $x^2 + y^2 = 63$
$y^2 - x^2 = 9$

17.________________

18. $x^2 + y^2 = 92$
$x^2 - y^2 = 70$

18.________________

19. $x^2 + 4y^2 = 21$
$x^2 - y^2 = 6$

19.________________

20. $8x^2 + 3y^2 = 39$
$y^2 - 2x^2 = 6$

20.________________

Name: Date:
Instructor: Section:

21. $x^2 + y^2 = 4$
$9x^2 + y^2 = 9$

21.________________

22. $x^2 + y^2 = 7$
$x^2 + 4y^2 = 16$

22.________________

23. $6x^2 + 2y^2 = 62$
$7x^2 + 4y^2 = 114$

23.________________

24. $3x^2 + 4y^2 = 47$
$5x^2 + 6y^2 = 75$

24.________________

Objective 3 Solve applications of nonlinear systems.

Solve.

25. A rectangle has a perimeter of 62 inches and a diagonal of 25 inches. Find the dimensions of the rectangle.

25.________________

Name: Date:
Instructor: Section:

26. A rectangle has a perimeter of 206 inches and a diagonal of 73 inches. Find the dimensions of the rectangle.

26.________________

27. A rectangular classroom has a perimeter of 128 feet and an area of 960 square feet. Find the dimensions of the classroom.

27.________________

28. A rectangle has a perimeter of 306 inches and a diagonal of 117 inches. Find the dimensions of the rectangle.

28.________________

29. A rectangle has a perimeter of 142 inches and a diagonal of 61 inches. Find the dimensions of the rectangle.

29.________________

30. A projectile is fired upward from the ground with an initial velocity of 76 feet per second. At the same instant, another projectile is fired upward from a roof 140 feet above the ground with an initial velocity of 41 feet per second.

a) At what time will the two projectiles be the same height above the ground?

b) How high above the ground are the two projectiles at that time?

30.a.________________

b.________________

Name: Date:
Instructor: Section:

Chapter 10 SEQUENCES, SERIES AND THE BINOMIAL THEOREM

10.1 Sequences and Series

Learning Objectives

1. Find the terms of a sequence, given its general term.
2. Find the general term of a sequence.
3. Find partial sums of a sequence.
4. Use summation notation to evaluate a series.
5. Use sequences to solve applied problems.

Key Terms

Use the most appropriate term or phrase from the given list to complete each statement in exercises 1-4.

sequence	**infinite**	**finite**	**general**	**alternating**
recursively	**series**	**upper**	**lower**	**index**

1. A sequence whose terms alternate between being positive and negative is called a(n) ______________________ sequence.

2. The domain of a(n) ______________________ sequence is the set $\{1, 2, 3, \ldots, n\}$ for some natural number n.

3. In the notation $\sum_{i=1}^{n} a_i$ the number 1 is called the ______________________ limit of summation.

4. An infinite ______________________ is the sum of the terms of an infinite sequence.

Objective 1 Find the terms of a sequence, given its general term.

1. Find the 7th term of the sequence whose general term is $a_n = 6n + 13$.

1.________________

Name: Date:
Instructor: Section:

2. Find the 8th term of the sequence whose general term is $a_n = \dfrac{1}{n+5}$. **2.**________________

Find the first five terms of the sequence with the given general term.

3. $a_n = \left(\dfrac{2}{3}\right)^n$ **3.**________________

4. $a_n = 20 \cdot (-0.1)^{n-1}$ **4.**________________

5. $a_n = -4n$ **5.**________________

6. $a_n = 2n + 15$ **6.**________________

Objective 2 Find the general term of a sequence.

Find the next three terms of the given sequence, and find its general term a_n.

7. $-1,\ 4,\ -16,\ 64,\ -256,\ ...$ **7.**________________

8. $\dfrac{11}{2},\ 6,\ \dfrac{13}{2},\ 7,\ \dfrac{15}{2},\ ...$ **8.**________________

9. $9,\ -16,\ 25,\ -36,\ 49,\ ...$ **9.**________________

Name: Date:
Instructor: Section:

10. $-\frac{1}{8}, \frac{1}{9}, -\frac{1}{10}, \frac{1}{11}, -\frac{1}{12}, \ldots$

10._______________

Find the first five terms of the sequence with the given first term and general term.

11. $a_1 = \frac{1}{2}$, $a_n = 3a_{n-1} + \frac{1}{4}$

11._______________

12. $a_1 = \frac{3}{4}$, $a_n = 5a_{n-1} - 2$

12._______________

13. $a_1 = 0.6$, $a_n = -2a_{n-1} + 0.5$

13._______________

14. $a_1 = 20$, $a_n = 3(a_{n-1})$

14._______________

Objective 3 Find partial sums of a sequence.

Find the indicated partial sum for the given sequence.

15. s_8, 16, 8, 4, 2, …

15._______________

16. s_{12}, $\frac{1}{25}, \frac{1}{5}$, 1, 5, …

16._______________

Name: Date:
Instructor: Section:

17. s_8, $-14, -6, 2, 10, \ldots$ **17.**________________

Find the indicated partial sum for the sequence with the given general term.

18. s_4, $a_n = 10n$ **18.**________________

19. s_{11}, $a_n = 2^n$ **19.**________________

Find the indicated partial sum for the sequence with the given first term and general term.

20. s_{13}, $a_1 = 3$, $a_n = -2a_{n-1}$ **20.**________________

21. s_9, $a_1 = -6$, $a_n = -4a_{n-1}$ **21.**________________

Objective 4 Use summation notation to evaluate a series.

Find the sum.

22. $\sum_{i=1}^{6} 2 \cdot 3^i$ **22.**________________

Name: Date:
Instructor: Section:

23. $\sum_{i=1}^{10}(i+6)$

23.________________

24. $\sum_{i=1}^{5}\frac{1}{i}$

24.________________

25. $\sum_{i=1}^{7}(-1)^i\cdot 3i$

25.________________

Rewrite the sum using summation notation.

26. $6+15+24+33$

26.________________

27. $8+3+(-2)+(-7)+(-12)$

27.________________

28. $1+\left(-\frac{1}{2}\right)+\frac{1}{4}+\left(-\frac{1}{8}\right)+\frac{1}{16}$

28.________________

29. $1+(-8)+27+(-64)$

29.________________

Name: Date:
Instructor: Section:

30. $0.4+0.04+0.004+0.0004$

30.________________

Objective 5 Use sequences to solve applied problems.

Solve.

31. A college established a scholarship program for incoming freshmen, with funds to help 200 freshmen the first year.

a. If the program is structured to grow by 25 students per year, construct a sequence showing the number of students served in each of the first seven years.

b. How many students will have been helped by the program in the first seven years?

31.a.________________

b.________________

32. A magazine had 100,000 subscribers in its first year. If the publisher plans to increase the number of subscribers by 10% each year, construct a sequence showing the number of subscribers in each of the first six years.

32.________________

Name: Date:
Instructor: Section:

Chapter 10 SEQUENCES, SERIES AND THE BINOMIAL THEOREM

10.2 Arithmetic Sequences and Series

Learning Objectives
1. Find the general term of an arithmetic sequence.
2. Find partial sums of an arithmetic sequence.
3. Use an arithmetic sequence to solve an applied problem.

Key Terms
Use the most appropriate term or phrase from the given list to complete each statement in exercises 1-3.

arithmetic **common** **general** **difference** **sequence**

1. The __________________ term of an arithmetic sequence can be defined recursively by $a_n = a_{n-1} + d$ for $n \neq 1$.

2. The __________________ difference of an arithmetic sequence is the difference between consecutive terms.

3. An arithmetic series is the sum of the terms of an arithmetic__________________.

Objective 1 Find the general term of an arithmetic sequence.

Find the common difference, d, of the given arithmetic sequence.

1. $7, 3, -1, -5, -9, \ldots$ **1.**______________

2. $\frac{7}{2}, \frac{11}{3}, \frac{23}{6}, 4, \frac{25}{6}, \ldots$ **2.**______________

3. $87, 68, 49, 30, 11, \ldots$ **3.**______________

4. $7, -16, -39, -62, -85, \ldots$ **4.**______________

Name: Date:
Instructor: Section:

5. $\frac{17}{6}, \frac{43}{12}, \frac{13}{3}, \frac{61}{12}, \frac{35}{6}, \ldots$ **5.**__________

6. $\frac{93}{4}, 18, \frac{51}{4}, \frac{15}{2}, \frac{9}{4}, \ldots$ **6.**__________

7. $\frac{3}{8}, \frac{19}{24}, \frac{29}{24}, \frac{13}{8}, \frac{49}{24}, \ldots$ **7.**__________

8. 52, 39, 26, 13, 0, … **8.**__________

9. 28, 23, 18, 13, 8, … **9.**__________

Find the first five terms of the arithmetic sequence with the given first term and common difference.

10. $a_1 = 6$, $d = -9$ **10.**__________

11. $a_1 = 7$, $d = 13$ **11.**__________

12. $a_1 = 0$, $d = -16$ **12.**__________

13. $a_1 = 78$, $d = 39$ **13.**__________

Name: Date:
Instructor: Section:

14. $a_1 = \frac{13}{2}$, $d = \frac{9}{4}$

14.________________

15. $a_1 = \frac{3}{8}$, $d = 6$

15.________________

Find the general term, a_n, of the given arithmetic sequence.

16. 5, 11, 17, 23, 29, ...

16.________________

17. 4.3, 3.2, 2.1, 1, -0.1, ...

17.________________

18. 3723, 2794, 1865, 936, 7, ...

18.________________

19. Find the 32nd term, a_{32}, of the arithmetic sequence

$-7, -13, -19, -25, -31, \ldots.$

19.________________

Name: Date:
Instructor: Section:

20. Find the 21st term, a_{21}, of the arithmetic sequence

20.________________

$139,\ 72,\ 5,\ -62,\ -129,\ \ldots.$

21. Find the 24th term, a_{24}, of the arithmetic sequence

21.________________

$21,\ \frac{105}{4},\ \frac{63}{2},\ \frac{147}{4},\ 42,\ \ldots.$

22. Find the 50th term, a_{50}, of the arithmetic sequence

22.________________

$-84,\ -34,\ 16,\ 66,\ 116,\ \ldots.$

23. Find the 26th term, a_{26}, of the arithmetic sequence

23.________________

$-14,\ -5,\ 4,\ 13,\ 22,\ \ldots.$

Name: Date:
Instructor: Section:

24. Which term of the arithmetic sequence

$-13, -6, 1, 8, 15, \ldots$ is equal to 274?

24.________________

25. Which term of the arithmetic sequence
$-3474, -3448, -3422, -3396, -3370, \ldots$
is equal to 2272?

25.________________

26. Which term of the arithmetic sequence

$-29, 51, 131, 211, 291, \ldots$ is equal to 32,291?

26.________________

Objective 2 Find partial sums of an arithmetic sequence.

Solve.

27. $s_{40}, \frac{17}{2}, 29, \frac{99}{2}, 70, \ldots$

27.________________

28. $s_{50}, \frac{5}{2}, \frac{25}{2}, \frac{45}{2}, \frac{65}{2}, \ldots$

28.________________

Name: Date:
Instructor: Section:

29. s_{16}, $-99,\ -213,\ -327,\ -441,\ \ldots$ **29.**________________

30. s_{12}, 5000, 7000, 9000, 11,000, ... **30.**________________

31. Find the sum of the first 125 odd positive integers. **31.**________________

Objective 3 Use an arithmetic sequence to solve an applied problem.

Solve.

32. An online music store sold 30,000 copies of a new CD on the first day of its release. The number of copies sold decreased by 2500 on each successive day. **32.a.**______________

a. Write a sequence showing the number of CDs sold on the first six days. **b.**______________

b. Find the general term a_n for the number of CDs sold on the *n*th day.

33. Tina made a stack of fence posts in such a way that each row in the stack has one less post than the row directly below it. The top row of the stack has 8 posts. **33.a.**______________

a. If the top row is called row 1, find the general term a_n for the number of posts in row n. **b.**______________

b. If the stack has a total of 225 posts, how many rows are there?

Name: Date:
Instructor: Section:

34. A copy machine was purchased for $8000. Its value decreases by $1250 each year. Write a sequence showing the value of the copy machine at the end of each of the first five years.

34.________________

35. Keith opened a new fast food restaurant. His goal is to generate sales of $7500 in the first month, and then increase sales at a rate of $1000 per month.

a. Write a sequence showing Keith's sales goals for the first 8 months.

b. Find the general term a_n for the sales goal of the nth month.

c. If Keith meets his goals, what will the total sales be for the first 2 years?

35.a.________________

b.________________

c.________________

Name: Date:
Instructor: Section:

Chapter 10 SEQUENCES, SERIES AND THE BINOMIAL THEOREM

10.3 Geometric Sequences and Series

Learning Objectives
1. Find the common ratio of a geometric sequence.
2. Find the general term of a geometric sequence.
3. Find partial sums of a geometric sequence.
4. Find the sum of an infinite geometric series.
5. Use a geometric sequence to solve an applied problem.

Key Terms

Use the most appropriate term or phrase from the given list to complete each statement in exercises 1-3.

geometric **ratio** **general** **partial** **sequence** **limit**

1. The ____________________ term of a geometric sequence can be defined recursively by $a_n = r \cdot a_{n-1}$ for $n \neq 1$.

2. The common __________________ between consecutive terms in a geometric sequence is given by $\frac{a_n}{a_{n-1}}$.

3. The sum of the first n terms of an infinite geometric sequence is called a __________________ sum.

Objective 1 Find the common ratio of a geometric sequence.

Find the common ratio, r, of the given geometric sequence.

1. $40, 10, \frac{5}{2}, \frac{5}{8}, \frac{5}{32}, \ldots$

1.________________

2. $35, 14, \frac{28}{5}, \frac{56}{25}, \frac{112}{125}, \ldots$

2.________________

Name: Date:
Instructor: Section:

3. $-\frac{135}{4}, 45, -60, 80, -\frac{320}{3},\ldots$ **3.**________________

4. $432, -72, 12, -2, \frac{1}{3}, \ldots$ **4.**________________

Find the first five terms of the geometric sequence with the given first term and common ratio.

5. $a_1 = \frac{1}{32}, r = -4$ **5.**________________

6. $a_1 = \frac{5}{48}, r = 8$ **6.**________________

7. $a_1 = 0.2, r = 0.1$ **7.**________________

8. $a_1 = 10, r = -4$ **8.**________________

Objective 2 Find the general term of a geometric sequence.

Find the general term, a_n, of the given geometric sequence.

9. $54, 18, 6, 2, \ldots$ **9.**________________

Name: Date:
Instructor: Section:

10. $-3.4,\ 4.08,\ -4.896,\ 5.8752,\ \ldots$

10.________________

11. Find the ninth term, a_9, of the geometric sequence $1, 9, 81, 729, \ldots.$

11.________________

12. Find the sixth term, a_6, of the geometric sequence $8, 56, 392, 2744, \ldots.$

12.________________

13. Find the tenth term, a_{10}, of the geometric sequence $48, 12, 3, \frac{3}{4}, \ldots.$

13.________________

14. Find the thirteenth term, a_{13}, of the geometric sequence $\frac{4}{3}, \frac{8}{9}, \frac{16}{27}, \frac{32}{54}, \ldots.$

14.________________

15. Find the eighth term, a_8, of the geometric sequence $25, -10, 4, -\frac{8}{5}, \ldots.$

15.________________

Name: Date:
Instructor: Section:

Objective 3 Find partial sums of a geometric sequence.

Find the indicated partial sum for the given geometric sequence.

16. s_9, 200, -20, 2, $-\frac{1}{5}$, ... **16.**______________

17. s_6, 11, $\frac{22}{5}$, $\frac{44}{25}$, $\frac{88}{125}$, ... **17.**______________

18. s_{17}, 2, 8, 32, 128, ... **18.**______________

Find the partial sum of the geometric sequence with the given first term and common ratio.

19. s_{14}, $a_1 = 10{,}240$, $r = \frac{1}{2}$ **19.**______________

20. s_5, $a_1 = 25$, $r = -\frac{4}{5}$ **20.**______________

Name: Date:
Instructor: Section:

Objective 4 Find the sum of an infinite geometric series.

For the given geometric sequence, does the infinite series have a limit? If so, find that limit.

21. 3, 2.7, 2.43, 2.187, … **21.**________________

22. 42, -12.6, 3.78, -1.134, … **22.**________________

23. 0.2, -0.02, 0.002, -0.0002, … **23.**________________

24. 0.04, 0.0004, 0.000004, 0.00000004, … **24.**________________

25. 42, 7, $\frac{7}{6}$, $\frac{7}{36}$, … **25.**________________

26. For the geometric sequence whose general term is $a_n = 10 \cdot \left(\frac{3}{8}\right)^{n-1}$, does the infinite series have a limit? If so, what is that limit? **26.**________________

Name: Date:
Instructor: Section:

Objective 5 Use a geometric sequence to solve an applied problem.

Solve.

27. A new catering truck was purchased for $30,000. The value of the truck decreases by 40% each year. Write a geometric sequence showing the value of the truck at the end of each of the first 4 years after it was purchased.

27.________________

Name: Date:
Instructor: Section:

Chapter 10 SEQUENCES, SERIES AND THE BINOMIAL THEOREM

10.4 The Binomial Theorem

Learning Objectives
1 Evaluate factorials.
2 Calculate binomial coefficients.
3 Use the binomial theorem to expand a binomial raised to a power.
4 Use Pascal's triangle and the binomial theorem to expand a binomial raised to a power.

Key Terms

Use the most appropriate term from the given list to complete each statement in exercises 1-2.

factorial **binomial** **expand** **coefficient**

1. The product of all the positive integers from 1 through some positive integer n is called n ________________.

2. For any two positive integers n and r, $n \geq r$, the number ${}_nC_r$ is a ________________ coefficient.

Objective 1 Evaluate factorials.

Evaluate.

1. 11 ! **1.**________________

2. 13 ! **2.**________________

3. 12 ! **3.**________________

4. 7 ! **4.**________________

5. 3 ! **5.**________________

6. 0 ! **6.**________________

Name: Date:
Instructor: Section:

Rewrite the given product as a factorial.

7. $1 \cdot 2 \cdot 3 \cdot \ldots \cdot 123$ **7.**________________

8. $1 \cdot 2 \cdot 3 \cdot \ldots \cdot 500$ **8.**________________

Simplify.

9. $\frac{13\,!}{12\,!}$ **9.**________________

10. $\frac{3\,!}{6\,!}$ **10.**________________

11. $\frac{10\,!}{1\,! \cdot 9\,!}$ **11.**________________

12. $\frac{10\,!}{5\,! \cdot 5\,!}$ **12.**________________

13. $9\,! - 3\,!$ **13.**________________

14. $6\,! \cdot 9\,!$ **14.**________________

15. $\frac{13\,!}{5\,! \cdot 8\,!}$ **15.**________________

Name: Date:
Instructor: Section:

Objective 2 Calculate binomial coefficients.

Evaluate the given binomial coefficient.

16. ${}_{12}C_7$ 16.________________

17. ${}_{8}C_3$ 17.________________

18. ${}_{15}C_2$ 18.________________

19. ${}_{9}C_1$ 19.________________

Objective 3 Use the binomial theorem to expand a binomial raised to a power.

Expand using the binomial theorem.

20. $(2x-y)^4$ 20.________________

21. $(x-y)^6$ 21.________________

22. $(x-y)^{10}$ 22.________________

Name: Date:
Instructor: Section:

23. $(5x + 2y)^2$

23.________________

24. $(3x - 4y)^5$

24.________________

25. $(8x - 7y)^7$

25.________________

26. $(3x - 1)^{10}$

26.________________

27. $(x - y)^5$

27.________________

28. $(3x + y)^4$

28.________________

Name: Date:
Instructor: Section:

29. $(x-5y)^5$ 29.________________

Objective 4 Use Pascal's triangle and the binomial theorem to expand a binomial raised to a power.

Expand using the binomial theorem and Pascal's triangle.

30. $(x+4y)^3$ 30.________________

31. $(3x+2y)^4$ 31.________________

32. $(5x-6y)^4$ 32.________________

33. $(6x-y)^5$ 33.________________

34. $(x-y)^7$ 34.________________

Name: Date:
Instructor: Section:

35. $(2x-y)^5$

35._______________

Name: Date:
Instructor: Section:

Appendix A SYNTHETIC DIVISION

Divide using synthetic division.

1. $\dfrac{x^2-6x+8}{x-4}$

1.________________

2. $\dfrac{x^2-8x+15}{x-5}$

2.________________

3. $\left(x^2-4x-32\right)\div(x+4)$

3.________________

4. $\left(x^2+x-6\right)\div(x+3)$

4.________________

5. $\dfrac{3x^2-x-3}{x-3}$

5.________________

Name: Date:
Instructor: Section:

6. $\dfrac{2x^2 - x - 2}{x - 5}$

6.________________

7. $\dfrac{3x^2 - x - 5}{x - 5}$

7.________________

8. $\dfrac{4x^2 - x - 4}{x - 4}$

8.________________

9. $\dfrac{2x^3 - 19x - 8}{x + 3}$

9.________________

10. $\dfrac{3x^3 - 51x - 13}{x + 4}$

10.________________

Chapter 1 REVIEW OF REAL NUMBERS

1.1 Integers, Opposites, and Absolute Value

Key Terms

1. opposite 2. whole numbers 3. greater than

4. negative numbers 5. quotient 6. exponent; base

Objective 1, Objective 2

1. $>$ 3. $>$

Objective 3

5. 3049 7. 4 9. 171 11. $-20, 20$

Objective 4

13. -17 15. 2 17. -15 19. -28

Objective 5

21. -840 23. 26 25. -115 27. 75

Objective 6

29. 100,000,000,000 31. -729 33. 1

Objective 7

35. $\frac{19}{3}$ 37. 11 39. 1

1.2 Introduction to Algebra

Key Terms

1. commutative 2. variable expression 3. associative

4. like terms

Objective 1

1. $\frac{6}{x}$ 3. $3(x-13)$ 5. $80c$ 7. $35+5(n-1)$

Objective 2

9. 45.53 11. 23 13. $-290,400$ 15. 0

Objective 3

17. $-17a$

Objective 4

19. $54x+63$ 21. $20a-15b+35c+30$

Objective 5

23a. 1 b. $9x$ c. 9 25a. 2 b. $13x$, $10z$ c. 13, 10

Objective 6

27. $3y+12$ 29. $21c+60$

1.3 Linear Equations and Absolute Value Equations

Key Terms

1. zero 2. contradiction 3. identity 4. solve 5. null set

Objective 1

1. $\left\{-\frac{5}{3}\right\}$ 3. $\{6\}$ 5. $\{-8\}$ 7. $\{24\}$

Objective 2

9. $\{3\}$ 11. $\{-1\}$ 13. $\left\{\frac{17}{6}\right\}$ 15. $\{6\}$

Objective 3

17. $\varnothing$ 19. $\mathbb{R}$

Objective 4

21. $y = \dfrac{4-3x}{2}$ 23. $s = \dfrac{P}{4}$ 25. $h = \dfrac{S - 2\pi r^2}{2\pi r}$

Objective 5

27. $\{-14,\ 4\}$ 29. $\left\{-\dfrac{17}{3}\right\}$ 31. $\{5,\ 10\}$ 33. $\{-26,\ 16\}$

35. $\left\{-\dfrac{2}{5},\ 12\right\}$ 37. $|x-7| = 0$, Answers will vary.

1.4 Problem Solving: Applications of Linear Equations

Key Terms

1. equilateral   2. consecutive   3. circumference

Objective 1, Objective 2

1. 12, 56

Objective 3

3. 15 cm   5. A: $110°$, B: $70°$ 7. $60°$

9. 6 blocks by 6 blocks, 144 inches

Objective 4

11. 54, 55, 56   13. 21, 23, 25   15. Two

Objective 5

17. 9 hours

Objective 6

19. \$21,214   21. 16 \$5 chips

1.5 Linear Inequalities and Absolute Value Inequalities

Key Terms

1. absolute value 2. closed 3 compound

Objective 1

1. $(-\infty, 2)$

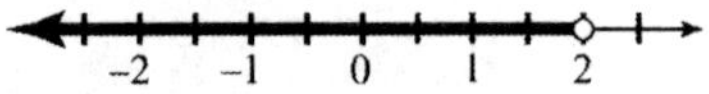

Objective 2

3. $(-\infty, 3]$

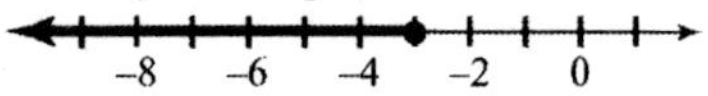

5. $(-3, \infty)$

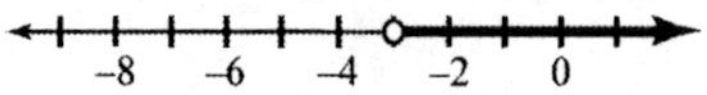

7. $[-5, \infty)$

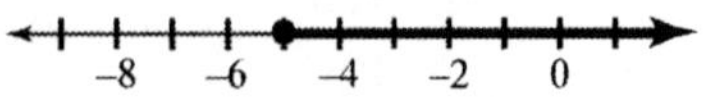

9. $\left(-\frac{11}{3}, \infty\right)$

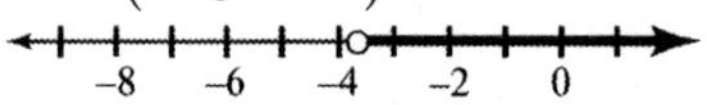

Objective 3

11. $(-1, 15)$

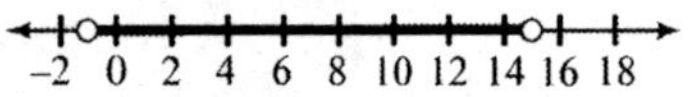

13. $(-2, 1]$

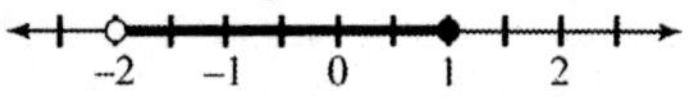

15. $\left[-3, \frac{1}{2}\right]$

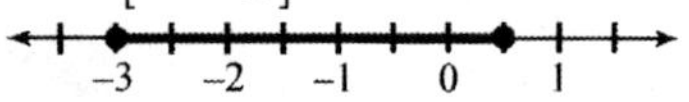

17. $(-\infty, -5] \cup \left[\frac{5}{4}, \infty\right)$

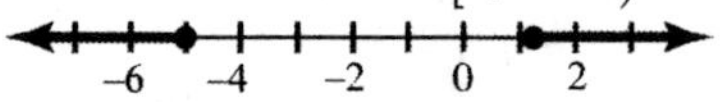

Objective 4

19. $x \geq 18$ 21. At least 1138 freshmen

Objective 5

23. $(-\infty, -4) \cup (4, \infty)$

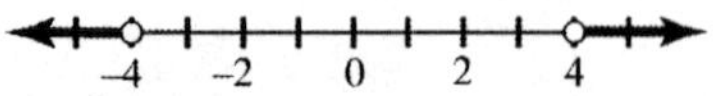

25. $\left(-\infty, -\frac{7}{4}\right) \cup \left(\frac{21}{4}, \infty\right)$

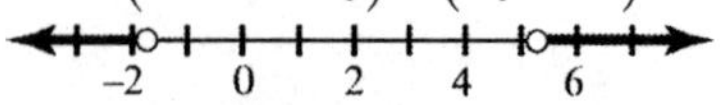

27. $\left(-\infty, \frac{1}{7}\right] \cup \left[\frac{3}{7}, \infty\right)$

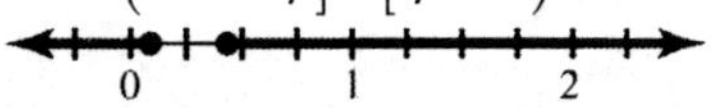

29. $[-8, 16]$

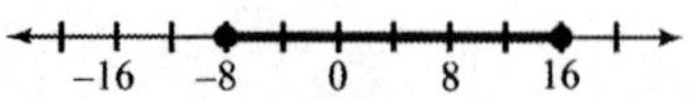

Chapter 2 GRAPHING LINEAR EQUATIONS

2.1 The Rectangular Coordinate System; Equations in Two Variables

Key Terms

1. Cartesian 2. *x*-axis 3. quadrants 4. standard

5. *y*; *x* 6. *x*-intercept

Objective 1

1. No 3. Yes

Objective 2

5.

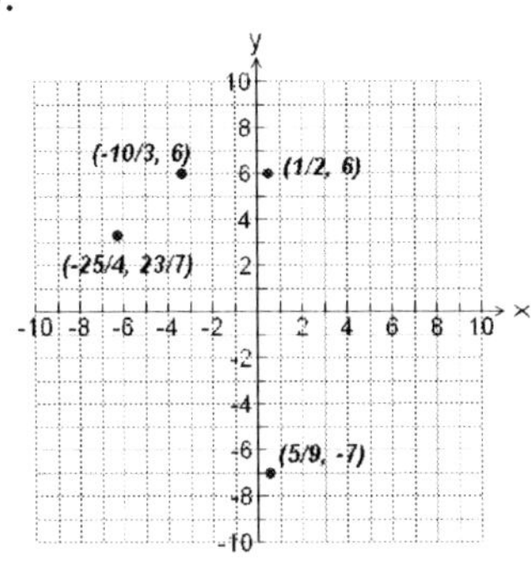

7.

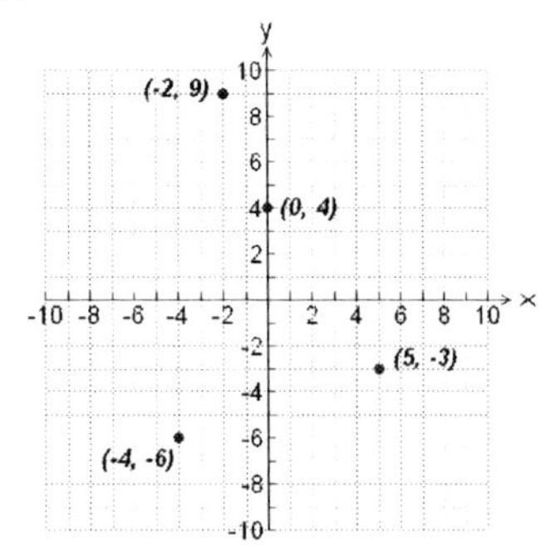

9. $A\ (0, 80)$, $B\ (-70,\ -90)$, $C\ (10,\ -60)$, $D\ (-50,\ 40)$

Objective 3

11. 6 13. 2

Objective 4

15. $(4,\ 0)$, No *y*-int. 17. No *x*-int., $(0,\ -3)$ 19. $(0,\ 0)$, $(0, 0)$

21. $(17,0)$, $(0,17)$

Objective 5

23. $\left(\frac{5}{2}, 0\right), \left(0, \frac{7}{2}\right)$

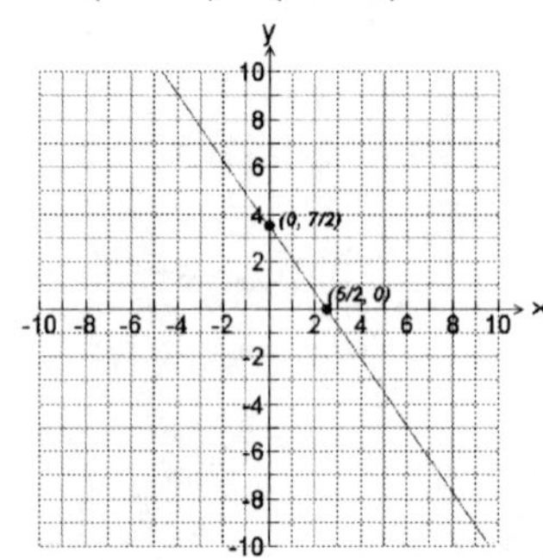

25. $(5,0), (0,2)$
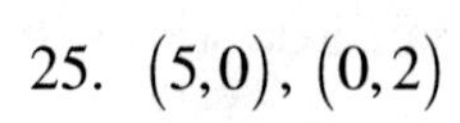

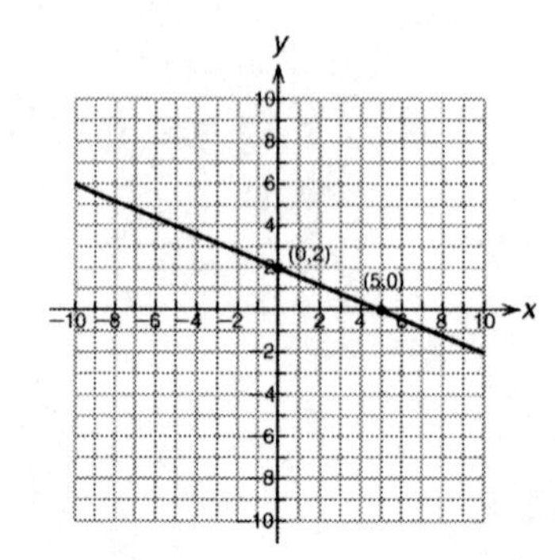

27. $(0,0), (0,0)$

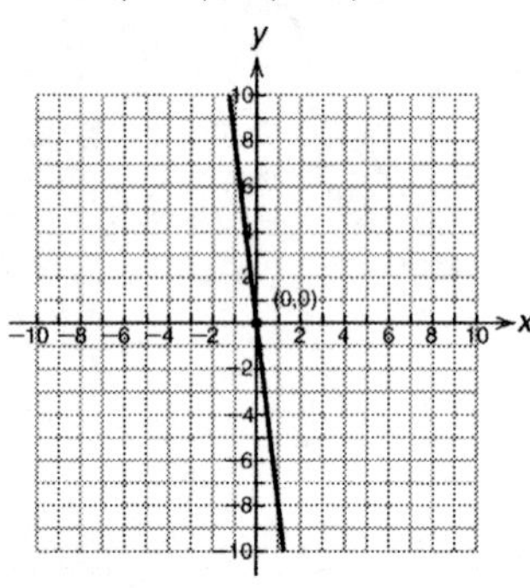

Objective 6

29. No x-int., $(0,-7)$

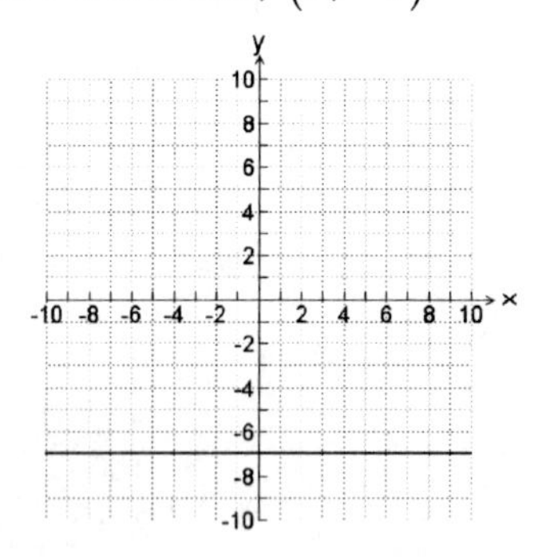

31. $(4,0)$, No y-int.

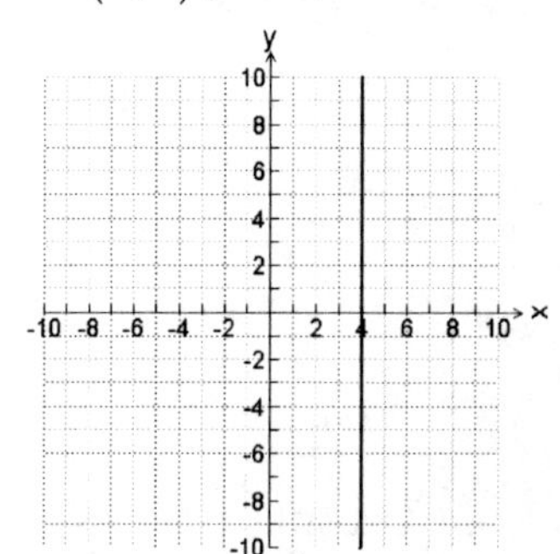

Objective 7

33a. $(0,0.39)$; Each call initially costs \$0.39.

b. The x-intercept is $(-13,0)$.

This answer does not make sense, since a person cannot make a negative number of calls.

2.2 Slope of a Line

Key Terms

1. second; first
2. undefined
3. parallel
4. negative reciprocals

Objective 1

1. Positive
3. $\frac{2}{3}$

Objective 2

5. $-\frac{2}{5}$
7. 1
9. Undefined
11. $\frac{1}{4}$

Objective 3

13. Undefined

Objective 4

15. 9, $(0,5)$
17. $\frac{6}{5}$, $(0,-3)$
19. $y=3$

Objective 5

21.

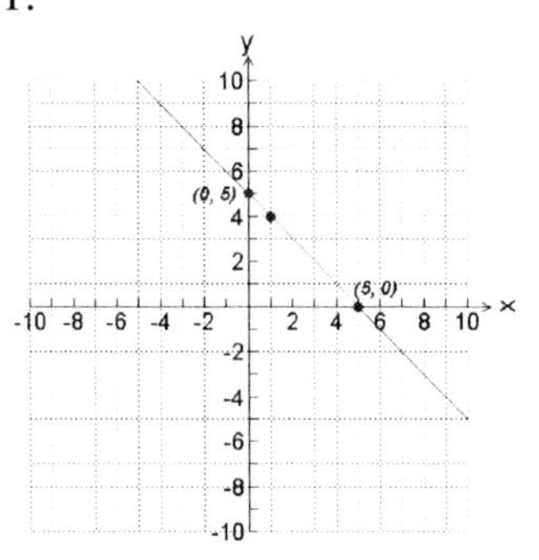

23.

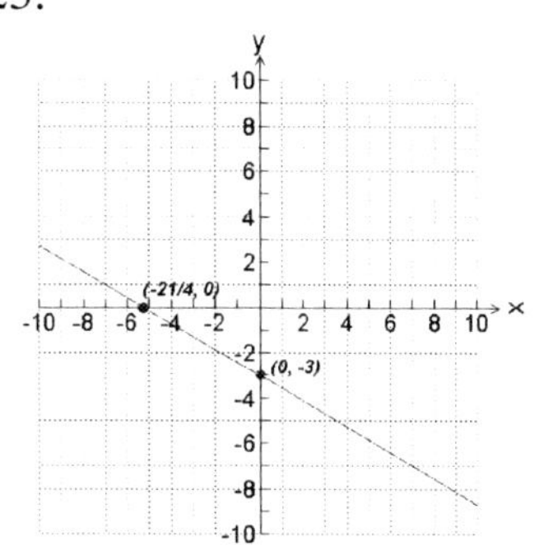

Objective 6

25. Yes
27. No
29. No
31. Yes
33. Neither
35. Perpendicular

Objective 7

37. $-\dfrac{3}{5}$

2.3 Equations of Lines

Key Terms

1. point-slope 2. slope

Objective 1

1. $y=-\dfrac{5}{3}x-\dfrac{7}{3}$ 3. $y=7x+5$ 5. $y=2x+1$ 7. $y=-x-5$

9. $y=x+9$

Objective 2

11. $y=-\dfrac{3}{4}x+8$

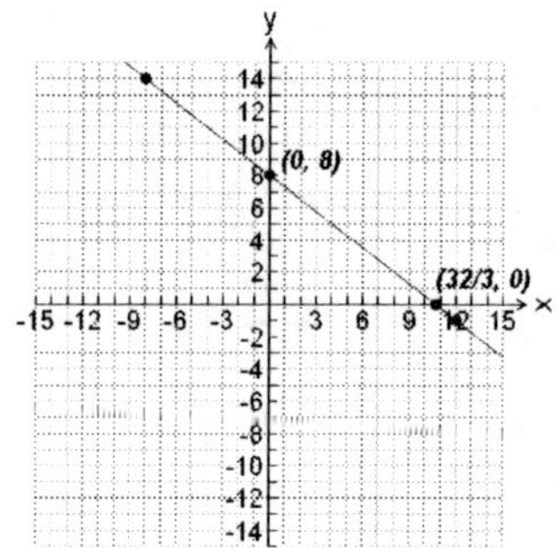

13. $y=\dfrac{2}{5}x-\dfrac{8}{5}$

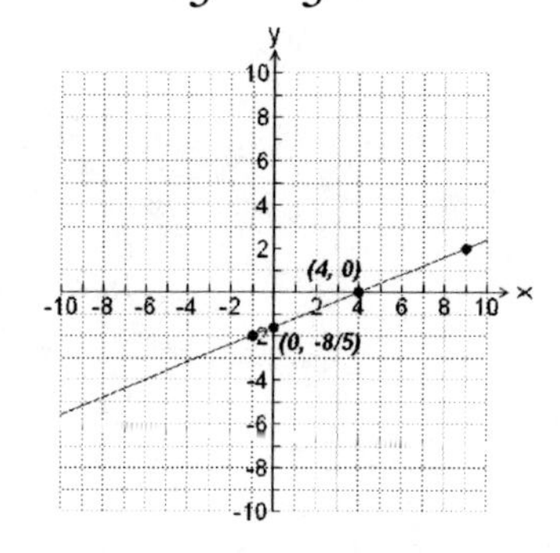

15. $y=-\dfrac{1}{2}x-2$

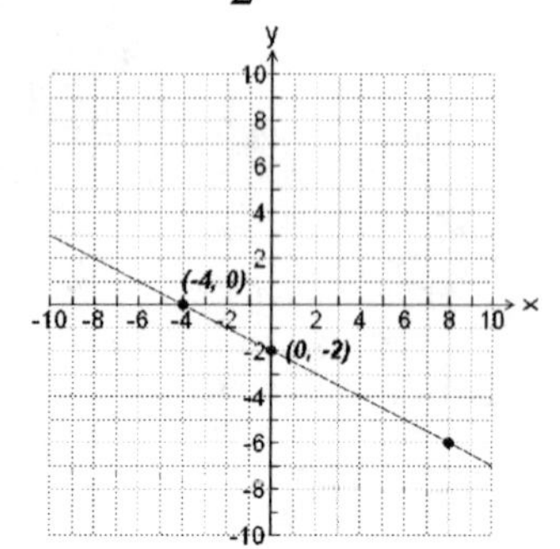

Objective 3

17a. $y = 30x + 100$ b. \$400

Objective 4

19. $x = -9$ 21. $y = 3$ 23. $y = -4$ 25. $y = x - 9$

2.4 Linear Inequalities

Key Terms

1. test point 2. strict

Objective 1

1. Yes 3. Yes

Objective 2

5. 7. A 9. >

11. 13.

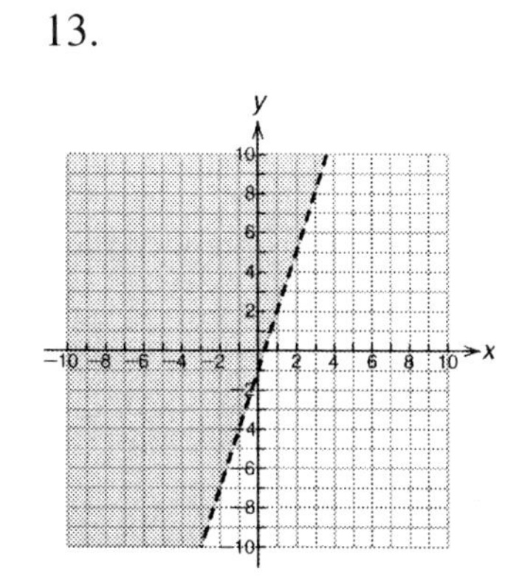

Objective 3

15.

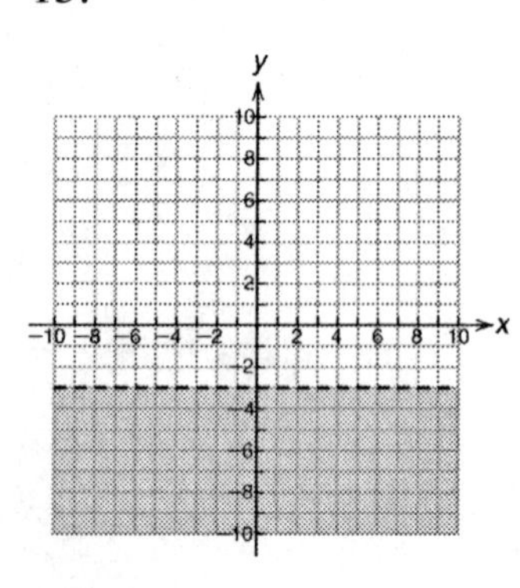

17.

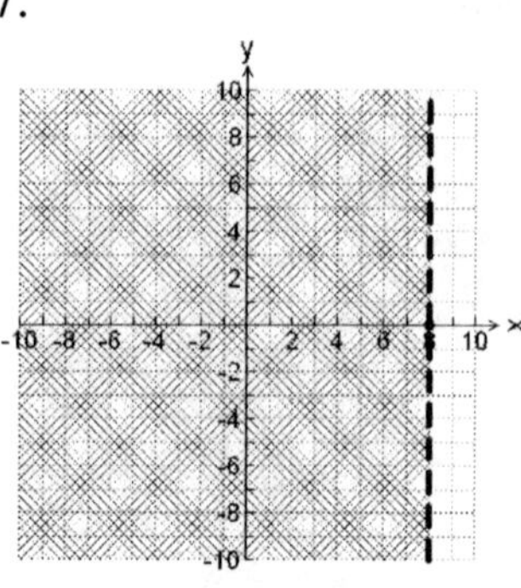

Objective 4

19a. $x + y \geq 1400$ b. $(700, 700)$ and $(1000, 500)$ 21. $x \leq -5$

2.5 Linear Functions

Key Terms

1. constant 2. relation 3. range

Objective 1

1. Yes; Yes 3. No, some x-coordinates are associated with more than one y-coordinate.

Objective 2

5. 11 7. 29

Objective 3

9.

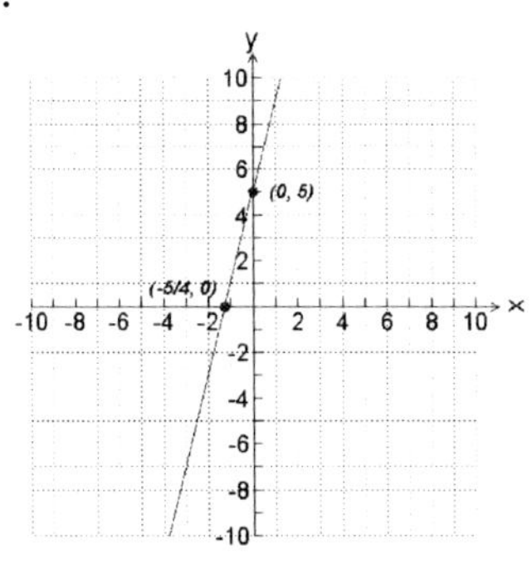

11.

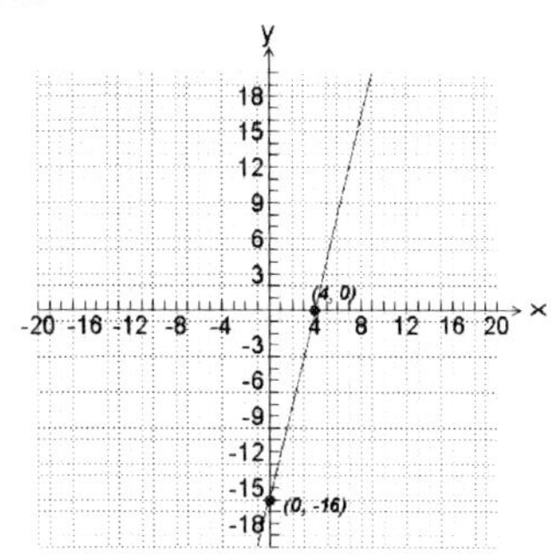

13.

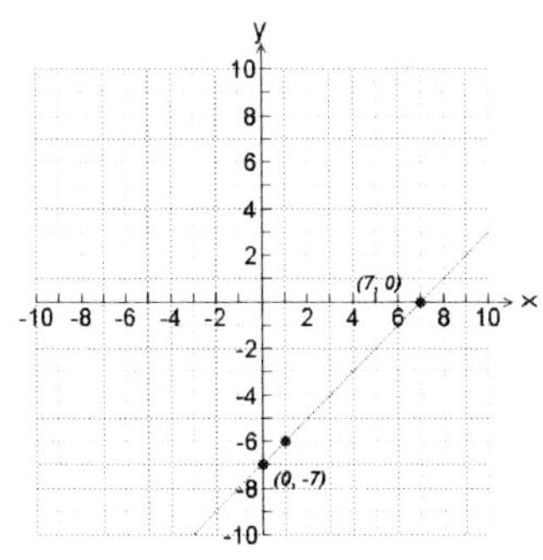

Objective 4

15a. $(-4,0)$ b. $(0,8)$ c. 2 d. -2

Objective 5

17. Domain: $(-\infty,\ \infty)$, Range: $(-\infty,\ \infty)$

19. Domain: $[-7,0]$, Range: $[-3,\ 8]$

Objective 6

21. $f(x)=3x+\dfrac{1}{2}$

23. $f(x)=\dfrac{1}{5}x-4$

25. $f(x)=-x+5$

2.6 Absolute Value Functions

Objective 1

1. Domain: $(-\infty, \infty)$, Range: $[6, \infty)$

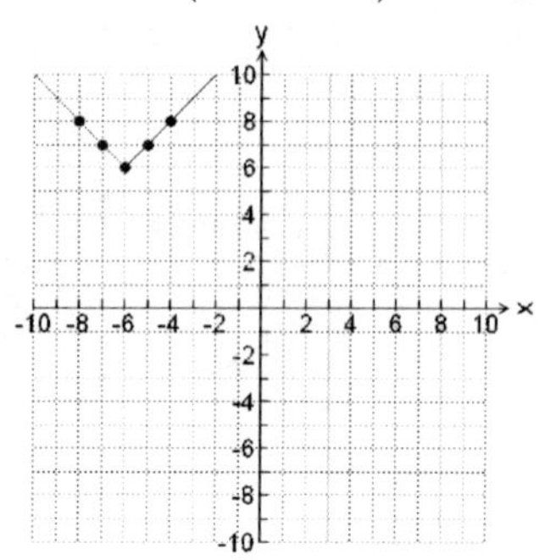

3. Domain: $(-\infty, \infty)$, Range: $[-2, \infty)$

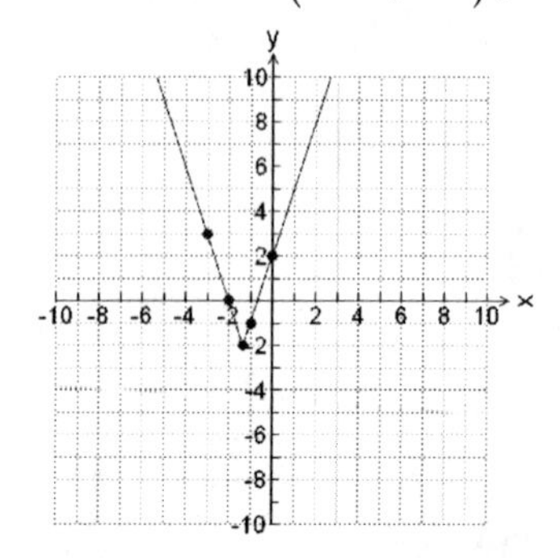

5. $(-3,0)$, $(9,0)$, $(0,-3)$ 7. $(4,0)$, $(0,4)$ 9. $[-9, 2]$

Objective 2

11. $f(x) = -|x-4| + 1$ 13. $f(x) = |x-5| - 3$ 15. $f(x) = -|x-2| - 4$

Chapter 3 SYSTEMS OF EQUATIONS

3.1 Systems of Two Linear Equations in Two Unknowns

Key Terms

1. inconsistent 2. dependent 3. independent

Objective 1

1. No 3. No

Objective 2

5. $\left(\frac{3}{2}, -\frac{7}{2}\right)$ 7. Dependent, $\left(x, \frac{4}{5}x+\frac{9}{5}\right)$ 9. $(0, 3)$

Objective 3

11. $(-1, -2)$ 13. $(6,1)$ 15. $(-3,3)$ 17. $(-7,-4)$

Objective 4

19. $(-7,1)$ 21. Inconsistent, $\varnothing$ 23. $\left(-\frac{34}{3}, \frac{58}{3}\right)$ 25. $(33, 18)$

3.2 Applications of Systems of Equations

Objective 1

1. 66 large, 29 medium   3. adult: \$3.50, child: \$1   5. 7 five dollar bills

7. won 50 games   9. 38 nickels, 40 quarters

Objective 2

11. 24°, 66°   13. 65.5°, 114.5°   15. 35°, 145°

17. length 42 meters, width 19 meters

Objective 3

19. \$5000 at 5.5%, \$9500 at 7%

Objective 4

21. 35 milliliters 8%, 21 milliliters 16%

23. 320 pounds cornmeal, 40 pounds soybean meal

Objective 5

25. Airplane: 150 mph, Wind: 30 mph   27. 100 mph

3.3 Systems of Linear Inequalities

Objective 1

1.

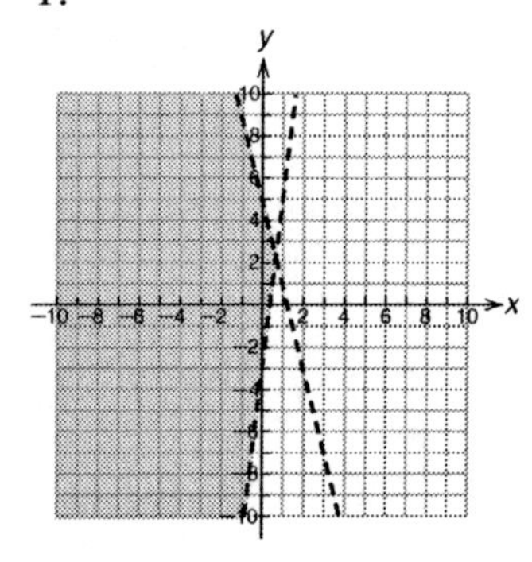

3.

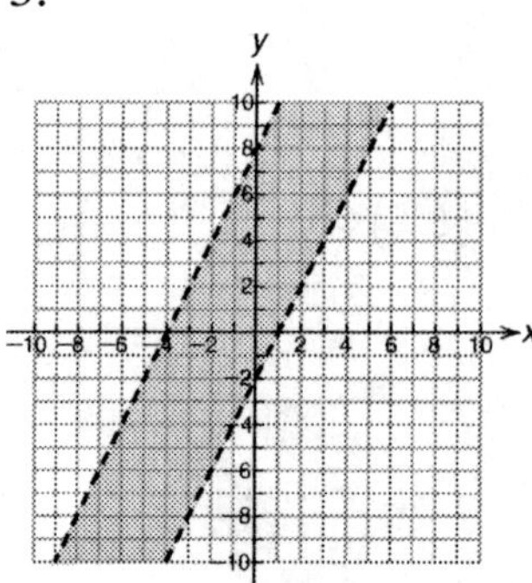

5.

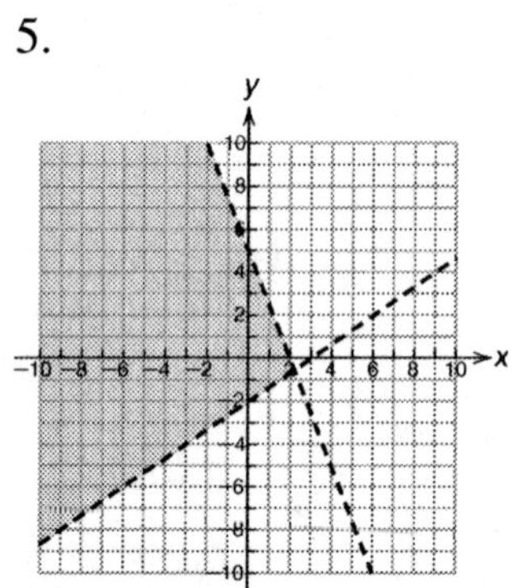

7.

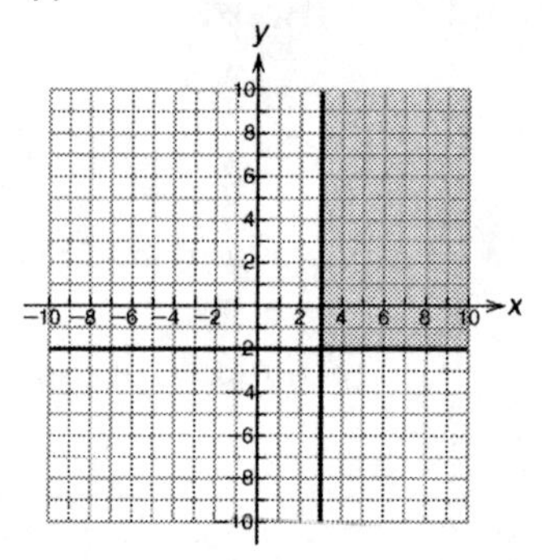

Objective 2

9.

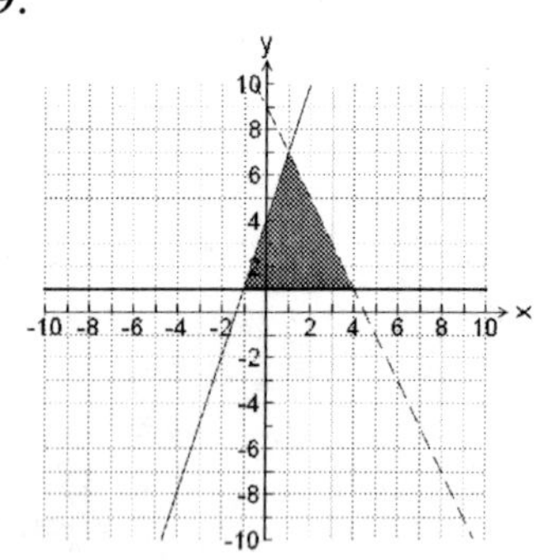

11.

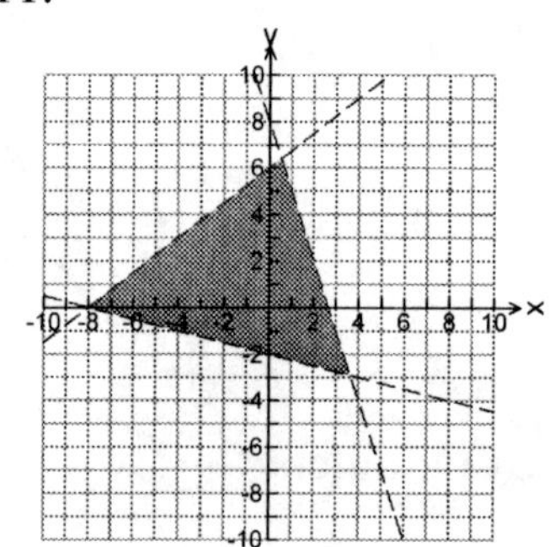

13.

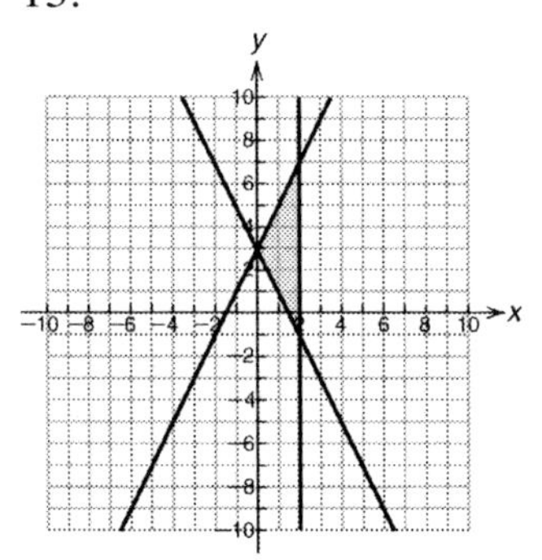

Objective 3

15. $x + y \leq 50$
$y \geq x + 30$

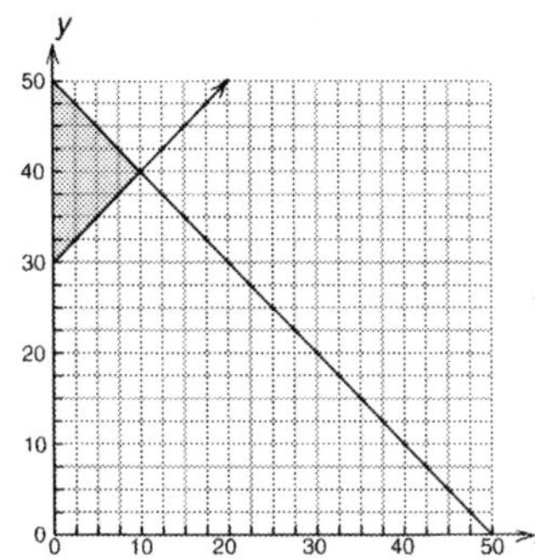

3.4 Systems of Three Equations in Three Unknowns

Objective 1

1. Yes 3. No

Objective 2

5. $(8, -6, -3)$ 7. $(3, -2, 0)$ 9. $(3, 3, -3)$ 11. $(5, -1, 5)$

Objective 3

13. $A: 21°$, $B: 86°$, $C: 73°$

3.5 Using Matrices to Solve Systems of Equations

Key Terms

1. rows; columns 2. augmented

Objective 1

1. $(-6, 8)$ 3. $(-7, 4)$ 5. $(-8, 9)$

Objective 2

7. $(-5, 2, -5)$ 9. $(-5, -3, 5)$

Objective 3

11. 266 attended the matinee, 357 attended the evening performance

13. 41 ones and 11 tens

3.6 Determinants and Cramer's Rule

Key Terms

1. square matrix 2. minor

Objective 1

1. 36 3. 15

Objective 2

5. $\left(\frac{82}{17}, \frac{47}{17}\right)$ 7. $\left(\frac{13}{2}, -\frac{9}{2}\right)$ 9. $(-9, 5)$

Objective 3

11. -8 13. -19

Objective 4

15. $(-6, 3, -3)$ 17. $(6, 5, 3)$

Objective 5

19. 83 cars and 136 trucks 21. \$36,000 at 8%, \$12,000 at 6%, \$20,000 at 9%

Chapter 4 EXPONENTS AND POLYNOMIALS

4.1 Exponents

Objective 1

1. x^{6y} 3. $-12x^{10}$ 5. 1 7. x^{3n} 9. $(a-b)^3$ 11. $10x^5$

13. -4 15. $\frac{216}{x^3}$

Objective 2

17. $432x^{22}$ 19. $\frac{4s^{24}t^{30}}{9x^{20}y^2}$ 21. 9 23. 4 25. 230

27. 3, 8

Objective 3

29. -9375

4.2 Negative Exponents; Scientific Notation

Objective 1

1. $\frac{1}{x^8}$ 3. $\frac{n^3}{m^3}$

Objective 2

5. $\frac{1}{x^{21}}$ 7. $\frac{9a^{10}c^8d^{14}}{b^{22}}$ 9. $\frac{y^{35}}{x^{10}}$ 11. $\frac{x^{18}y^{24}}{64z^{36}}$ 13. x^{35}

Objective 3

15. 9.23×10^7 17. 3.6×10^{-5}

Objective 4

19. 0.00000332 21. 800,000,000,000

Objective 5

23. 4.821×10^5 25. 3.4×10^{15}

Objective 6

27. 3.735×10^{-8} grams 29. 2.232×10^8 miles

4.3 Polynomials; Addition, Subtraction, and Multiplication of Polynomials

Objective 1

1. 3, 5, 2, 1 3. 5, 4, 3, 2, 0 5. -1, 8, 11 7. monomial

Objective 2

9. 0 11. 751 13. 52

Objective 3

15. $-18b^2-3b+10$ 17. $3x^6+4x^5-7x^4+x^3+9x-29$

19. $(f+g)(x)=8x^5+2x^4-13x^2-13x+35$,

$(f-g)(x)=-2x^4-12x^3-13x^2+13x+35$

Objective 4

21. $-63x^9$ 23. $-162x^{21}$ 25. $8xy^2z^4$ 27 $x^{11}-7x^{10}-30x^9$

29. $7x^5-5x^4-9x^3$ 31. x^4+17x^2+52 33. $4x$ 35. $-143x^{22}$

37. $x^2-21x+108$

Objective 5

39. x^2-100 41. $9x^2+12x+4$

Objective 6

43. 6, 8, 10, Polynomial: 10 45. $14rf+9r^2f+8rf^2$

4.4 An Introduction to Factoring; The Greatest Common Factor; Factoring by Grouping

Key Terms

1. term 2. binomial 3. product

Objective 1

1. 14

Objective 2

3. m^7n^5 5. $4x^4$

Objective 3

7. $7x^4y^3(3x^4+y)$ 9. $2x(7x^4-1)$ 11. $7a(3+5a-4a^2)$

Objective 4

13. $(6x+7)(x-1)$ 15. $(3x^2-5x)(7x+9)$

Objective 5

17. $(x+5)(x-8)$ 19. $(x+13)(2x-3)$ 21. $(2x+5)(3x^2+5)$

23. $(3x-8)(4x-7)$ 25. $(x+7)(x+2)(x-2)$

4.5 Factoring Trinomials of Degree 2

Key Terms

1. prime 2. perfect

Objective 1

1. $(x+6)(x+12)$ 3. $(x-6)(x-7)$ 5. $(x+15)(x-5)$

7. $(x+12)(x-11)$

Objective 2

9. $7x^{10}(x-7)(x+5)$ 11. $-x^4(x+2)(x+23)$ 13. $9x^4(x+2)^2$

15. $2(x^2-7x-24)$

Objective 3, Objective 4

17. $(12x-5)(x-4)$ 19. $2(x-5)(x+3)$ 21. $(3x-8)(4x+7)$

23. $(3x+7)(4x-9)$ 25. prime

Objective 5

27. $(x-3)^2$ 29. $(x+4)^2$

Objective 6

31. $(x-3y)(x-20y)$ 33. $(xy-10)(xy+3)$

4.6 Factoring Special Binomials

Objective 1

1. $(6+x)(6-x)$ 3. $-12(x^2+4y^2)$ 5. $10(x+9)(x-9)$

7. $(x^4+5)(x^4-5)$ 9. $(10+x^2)(10-x^2)$

Objective 2

11. $(x-1)(x+1)(x^2+1)(x^4+1)$ 13. $(x^4+y^4)(x^2+y^2)(x+y)(x-y)$

Objective 3

15. $(a^7-5b^5)(a^{14}+5a^7b^5+25b^{10})$ 17. $(2x-5y)(4x^2+10xy+25y^2)$

Objective 4

19. $x^2(y+5)(y^2-5y+25)$ 21. $(9+z)(81-9z+z^2)$

23. $(10x^{10}+9y^2z^7)(100x^{20}-90x^{10}y^2z^7+81y^4z^{14})$ 25. $-3(x+7)(x^2-7x+49)$

4.7 Factoring Polynomials: A General Strategy

Objective 1

1. $(2x-3y)(4x^2+6xy+9y^2)$ 3. $(x-4)(3x+1)(3x-1)$

5. $(x+4)(x+5)$ 7. $(x+4)(x+9)$ 9. $(x+6)(x+9)$

11. $(3x+4)^2$ 13. $(x+9)(x-7)$ 15. $(2x+5)(x-3)$

17. $(x-4)(4x+5)(4x-5)$ 19. $(x+12)(x-18)$ 21. $4(x-3)(x+6)$

23. $(x^2+5)(x^4-5x^2+25)$ 25. $(x+18)(x-22)$ 27. $(x+8)^2$

29. $4(3x-2)(x+4)$

4.8 Solving Quadratic Equations By Factoring

Key Terms

1. standard 2. factors 3. zero

Objective 1

1. $\{-4, 5\}$ 3. $\left\{2, -5, \frac{3}{4}\right\}$ 5. $\left\{-\frac{7}{3}, -6\right\}$

Objective 2

7. $\{1, 24\}$ 9. $\{-3, 3\}$

Objective 3

11. $\{4, -8\}$ 13. $\{-3, 10\}$ 15. $\{8\}$

Objective 4

17. $\left\{-\frac{5}{2}, 4\right\}$ 19. $\left\{-\frac{2}{5}, 1\right\}$

Objective 5

21. $x^2 - 11x + 24 = 0$ 23. $x^2 + 3x = 0$ 25. $9x^2 - 60x + 100 = 0$

27. $25x^2 - 1 = 0$ 29. $x = 9$

Objective 6

31. 9, 11 33. 6 inches by 12 inches

Chapter 5 RATIONAL EXPRESSIONS AND EQUATIONS

5.1 Rational Expressions and Functions

Key Terms

1. defined 2. numerator, denominator 3. rational

Objective 1

1. $\frac{5}{6}$ 3. $\frac{21}{17}$ 5. $\frac{31}{169}$

Objective 2

7. $-\frac{4}{7}$ 9. $-9,\ 4$ 11. $-8,\ 8$

Objective 3

13. $\frac{x+11}{x+3}$ 15. $\frac{3}{4x^2}$ 17. $\frac{x-5}{x+8}$

Objective 4

19. not opposites 21. opposites 23. $-\frac{8}{x-9}$

Objective 5

25. 10

Objective 6

27. $x \neq -3, 5$ 29. $x \neq 4, 6$ 31a. 8 b. 0, -3

5.2 Multiplication and Division of Rational Expressions

Objective 1

1. $\frac{x+1}{(x-3)(x-2)}$ 3. $\frac{(x-6)(x-2)}{(x+2)(x+6)}$ 5. $\frac{x+5}{(x-2)(x-3)}$

Objective 2

7. $\frac{x-7}{x+7}$ 9. $\frac{-x}{x-3}$ or $\frac{x}{3-x}$

Objective 3

11. $\frac{x-2}{2x+1}$ 13. $\frac{x-7}{x+7}$ 15. $\frac{(x+3)(x-2)}{x+1}$

Objective 4

17. $\frac{x-1}{(x-2)^2}$ 19. $\frac{5-x}{x+2}$

5.3 Addition and Subtraction of Rational Expressions

Objective 1

1. $\frac{13}{x-5}$ 3. $\frac{x+6}{x+2}$ 5. $\frac{(x+30)(x-2)}{(x+8)(x-3)}$ 7. $\frac{x-6}{x+5}$ 9. $\frac{11}{x}$

Objective 2

11. $-\frac{3}{x+10}$ 13. 4 15. $\frac{x+24}{x-10}$ 17. $-\frac{9}{x+55}$

Objective 3

19. $\frac{2x+7}{x-5}$ 21. $\frac{3x+1}{x-2}$ 23. $\frac{x+4}{4}$

Objective 4

25. $(x+2)(x-2)(x+8)$ 27. $(x+4)(x+4)(x+3)$

Objective 5

29. $\frac{23x}{15}$ 31. $\frac{6s+5r}{s^2r^2}$ 33. $\frac{3(x-24)}{(x+8)(x-8)}$ 35. $\frac{2(7x+8)}{(x+4)^2(x-4)}$

37. $x+6$ 39. $\frac{x-12}{(x+4)(x-4)}$

5.4 Complex Fractions

Objective 1

1. $\frac{25}{72}$ 3. $\frac{34}{7}$

Objective 2

5. $\frac{x+9}{x-8}$ 7. $\frac{2(4x-1)}{8x+3}$ 9. $\frac{x-2}{x}$ 11. $\frac{x+2}{x-2}$ 13. $\frac{x-4}{x-5}$

15. $-\frac{5}{2}$ 17. $\frac{x-7}{x+2}$ 19. $\frac{2}{x-5}$ 21. $\frac{x-1}{(x-6)(x+4)}$

23. $\frac{x-7}{x+8}$

5.5 Rational Equations

Key Terms

1. variables 2. undefined 3. denominators

Objective 1

1. $\left\{-\frac{480}{37}\right\}$ 3. $\left\{-\frac{12}{47}\right\}$ 5. $\{-5, 5\}$ 7. $\{-3\}$ 9. $\{-10\}$

11. $\{-8, 5\}$ 13. $\{-44\}$ 15. $\{-20, -2\}$ 17. $\{6, 12\}$

19. $\{-6, 7\}$

Objective 2

21. $c = \dfrac{ab}{a+b}$ 23. $M = \dfrac{5G - 5Ds}{D}$

5.6 Applications of Rational Equations

Key Terms

1. rate 2. inversely 3. directly

Objective 1

1. 6 3. 4, 12

Objective 2

5. 12 hours 7. 14 hours 9. 20 minutes

Objective 3

11. 125 km/hr

Objective 4

13. 21 grams 15. 75 mph 17. 400 feet

5.7 Division of Polynomials

Objective 1

1. $\dfrac{8}{x}$ 3. $23r^6$ 5. $28x^4y^4$

Objective 2

7. $7x^2 - 6x + 4$ 9. $x^3 - 5x^2 + 7x + 9$ 11. $-8x^8 + 20x^6 + 44x^5$

Objective 3

13. $3x - 5 - \dfrac{20}{4x+3}$ 15. $x + 2$ 17. $x^2 - 15x + 57$

Objective 4

19. $x^2+6x+27+\frac{191}{x-6}$ 21. $x^2-4x+16$

Chapter 6 RADICAL EXPRESSIONS AND EQUATIONS

6.1 Square Roots; Radical Notation

Key Terms

1. radical 2. root

Objective 1

1. 6 3. $\frac{5}{7}$

Objective 2

5. $11|x^7|$ 7. $m^{16}n^{18}$

Objective 3

9. 153.248 11. 9.950

Objective 4

13. x^5 15. 6 17. $-2x^2y^5z$

Objective 5

19. 9 21. 64

Objective 6

23. 3 25. 5 27. 54

Objective 7

29. 6.164

Objective 8

31. $(-\infty, 2]$ 33. $[8, \infty)$

6.2 Rational Exponents

Objective 1

1. $b^{1/3}$ 3. $7^{1/2}$ 5. $x^3y^2z^7$ 7. $9a^{17}b^{11}c^4$ 9. 5

Objective 2

11. $(2x^2y^3)^{3/4}$ 13. $m^{9/2}$ 15. 100,000 17. $100{,}000x^{55}y^{30}z^5$

19. 8 21. $128x^{49}y^{35}$

Objective 3

23. $x^{19/15}y^{3/2}$ 25. $8x^{8/15}$ 27. $x^{5/4}$ 29. $x^{23/24}$ 31. 1

33. $\frac{1}{2}$

Objective 4

35. $\frac{1}{2187}$ 37. $\frac{1}{2}$

Objective 5

39. $\sqrt[30]{w}$

6.3 Simplifying, Adding, and Subtracting Radical Expressions

Objective 1

1. $3\sqrt{5}$ 3. $3\sqrt[3]{6}$ 5. $x^{13}y^5z^{15}\sqrt[3]{xy^2}$ 7. $5x^5z^6\sqrt[3]{2yz^2}$

9. $3a^2b^3\sqrt[3]{15bc^2}$

Objective 2

11. $20\sqrt[4]{x^3}$ 13. $-13\sqrt{22}+13\sqrt[3]{22}$ 15. $19a\sqrt{n}-11\sqrt{m}$

Objective 3

17. $71a\sqrt{5}$ 19. $82\sqrt{7}+81\sqrt{3}-40$ 21. $58\sqrt{3}+25\sqrt{2}-14\sqrt{5}$

23. $-120\sqrt{7}+152\sqrt{3}$ 25. $4\sqrt{6}-10\sqrt{10}$

6.4 Multiplying and Dividing Radical Expressions

Key Terms

1. denominator 2. conjugate 3. conjugate

Objective 1

1. $-240\sqrt{3}$ 3. $210\sqrt[3]{14}$ 5. $14x^5\sqrt{15}$

Objective 2

7. $3\sqrt[3]{13}+3\sqrt[3]{15}$ 9. $-70\sqrt{3}-910$

Objective 3

11. -304 13. $8-2\sqrt{7}$ 15. $456+60\sqrt{3}$

Objective 4

17. -4 19. 4

Objective 5

21. 2 23. $2\sqrt[4]{3}$ 25. $\frac{\sqrt{6}}{x^7y^4}$ 27. $\frac{\sqrt{6}}{2}$ 29. $\frac{3\sqrt[3]{98}}{14}$

Objective 6

31. $\frac{28\sqrt{3}-18\sqrt{2}}{71}$ 33. $\frac{7\sqrt{2}-\sqrt{7}+5\sqrt{14}-5}{13}$ 35. $\frac{2\sqrt{3}+2\sqrt{11}}{3}$

37. $\dfrac{3\left(\sqrt{22}-2\sqrt{3}\right)}{2}$

6.5 Radical Equations and Applications of Radical Equations

Objective 1

1. $\varnothing$ 3. $\{162\}$ 5. $\{-1\}$

Objective 2

7. 4 9. -4, 8

Objective 3

11. $\{81\}$

Objective 4

13. $\{4, 9\}$ 15. $\{4\}$ 17. $\{-1\}$

Objective 5

19. $\{1, 9\}$ 21. $\{4\}$ 23. $\{3\}$

Objective 6

25. 13.0 feet 27. 3.51 seconds 29. 3.35 seconds 31. 4.49 seconds

Objective 7

33. 126 feet 35. 7.7 feet 37. 110.2 gallons/minute 39. 668 mph

6.6 The Complex Numbers

Key Terms

1. *a* 2. *b* 3. *i*

Objective 1

1. $3\sqrt{14}\,i$ 3. $8i$

Objective 2

5. $-1-17i$ 7. $33+8i$ 9. $5-47i$

Objective 3

11. -45 13. -54 15. $4i \cdot 8i$, Answers may vary.

Objective 4

17. $60-192i$ 19. $-119+120i$ 21. $-119-120i$ 23. 85

Objective 5

25. $\dfrac{63+22i}{73}$ or $\dfrac{63}{73}+\dfrac{22}{73}i$

Objective 6

27. $\dfrac{15-10i}{7}$ or $\dfrac{15}{7}-\dfrac{10}{7}i$ 29. $\dfrac{3-i}{2}$ or $\dfrac{3}{2}-\dfrac{1}{2}i$ 31. $15-8i$

Objective 7

33. 1 35. 0

Chapter 7 QUADRATIC EQUATIONS

7.1 Solving Quadratic Equations by Extracting Square Roots; Completing the Square

Key Terms

1. squared 2. completing the square

Objective 1

1. $\{-2, 6\}$ 3. $\{-6, 6\}$ 5. $\{-9, -7\}$

Objective 2

7. $\{-5, 5\}$ 9. $\{\pm 2\sqrt{5}\}$

Objective 3

11. $\left\{\frac{3\pm2\sqrt{6}}{2}\right\}$ 13. $\{-13, 5\}$ 15. $\left\{-\frac{7}{2}, \frac{5}{2}\right\}$

Objective 4

17. 8.5 meters 19. 6.0 inches

Objective 5

21. $x^2+20x+100=(x+10)^2$ 23. $x^2-\frac{7}{2}x+\frac{49}{16}=\left(x-\frac{7}{4}\right)^2$ 25. $\{5, 7\}$

27. $\left\{\frac{-2\pm3\sqrt{6}}{5}\right\}$ 29. $\{-8, -6\}$

Objective 6

31. $x^2-4x-1=0$ 33. $x^2+2x=0$ 35. $2x^2-5x-3=0$

37. $x^2-8x+13=0$

7.2 The Quadratic Formula

Key Terms

1. zero 2. negative 3. positive

Objective 1, Objective 2, Objective 3

1. $\left\{\frac{-13\pm3\sqrt{5}}{2}\right\}$ 3. $\left\{4\pm\sqrt{33}\right\}$ 5. $\left\{-\frac{1}{4}\right\}$ 7. $\{2, 6\}$

9. $\left\{\frac{-9\pm\sqrt{19}\,i}{2}\right\}$ 11. $\{-11, -2\}$

Objective 4

13. Two complex 15. Two real 17. Two complex 19. Two real

21. One real

Objective 5

23. Factorable   25. Prime   27. Factorable   29. Prime

31. $\left\{9 \pm 2\sqrt{6}\right\}$ 33. $\{-4, 8\}$ 35. $\left\{\frac{5}{3}, \frac{3}{2}\right\}$

Objective 6

37. 8.4 seconds

7.3 Equations That Are Quadratic in Form

Objective 1

1. $\{\pm i, -3, 3\}$ 3. $\left\{\pm\sqrt{2}, \pm 2\sqrt{3}\right\}$ 5. $\{36\}$ 7. $\{49\}$

9. $\{-8, 1331\}$ 11. $\{-1, -4, 2\}$

Objective 2

13. $\{24\}$ 15. $\{-7, 3\}$ 17. $\{-3, 9\}$ 19. $\{5\}$

Objective 3

21. $\{-6, 8\}$ 23. $\left\{-\frac{5}{2}, 1\right\}$ 25. $\{-10, 6\}$ 27. $\{-2, 8\}$

Objective 4

29. 45.5 hours

7.4 Graphing Quadratic Equations and Quadratic Functions

Key Terms

1. symmetric   2. axis of symmetry   3. x-intercepts

Objective 1

1. $(2, 11)$, $x = 2$ 3. $\left(-\frac{7}{8}, \frac{113}{16}\right)$, $x = -\frac{7}{8}$ 5. $\left(\frac{9}{2}, -\frac{125}{4}\right)$, $x = \frac{9}{2}$

7. $(-8,\ 0),\ (2,\ 0),\ (0,\ -16)$

9. $(0.8,\ 0),\ (5.2,\ 0),\ (0,\ -4)$

11. No x-intercepts, $(0,\ 52)$

13. Vertex: $\left(\frac{5}{2},\ -\frac{121}{4}\right)$, y-int: $(0,\ -24)$, x-int: $(-3,\ 0)$, $(8,\ 0)$

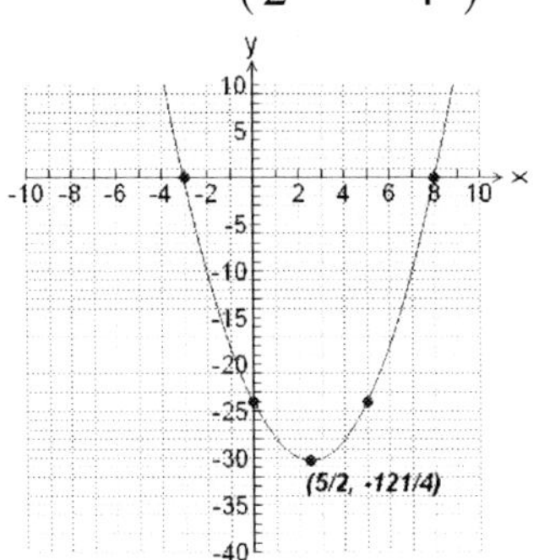

Objective 2

15. Vertex: $\left(\frac{3}{2},\ -\frac{7}{4}\right)$, y-int: $(0,\ -4)$, x-int: None

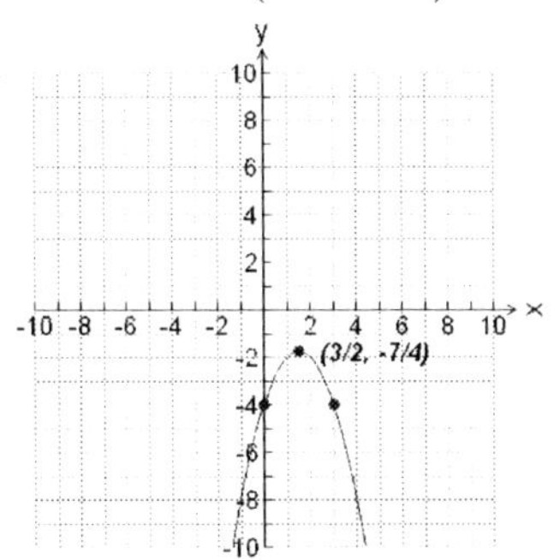

17. Vertex: $\left(\frac{5}{2},\ \frac{49}{4}\right)$, y-int: $(0,\ 6)$, x-int: $(-1,\ 0)$, $(6,\ 0)$

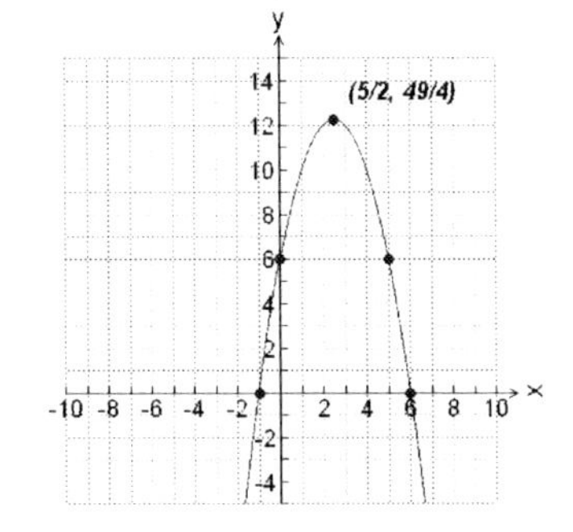

Objective 3

19. Vertex: $(1,\ 5)$, y-int: $(0,\ 6)$, x-int: None

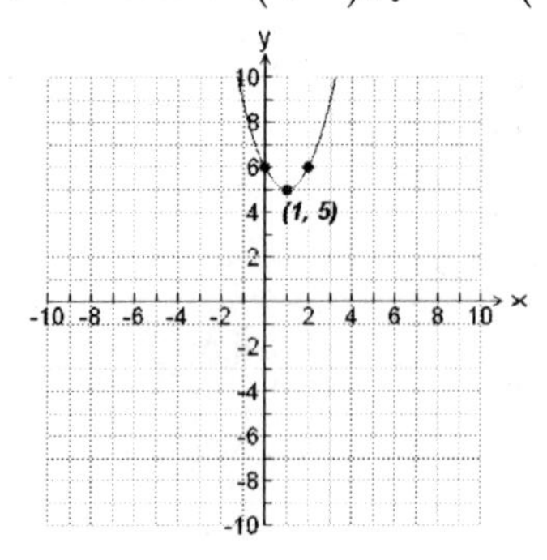

21. Vertex: $\left(-\frac{5}{2},\ \frac{37}{4}\right)$, y-int: $(0,\ 3)$, x-int: $\left(\frac{-5\pm\sqrt{37}}{2},\ 0\right)$

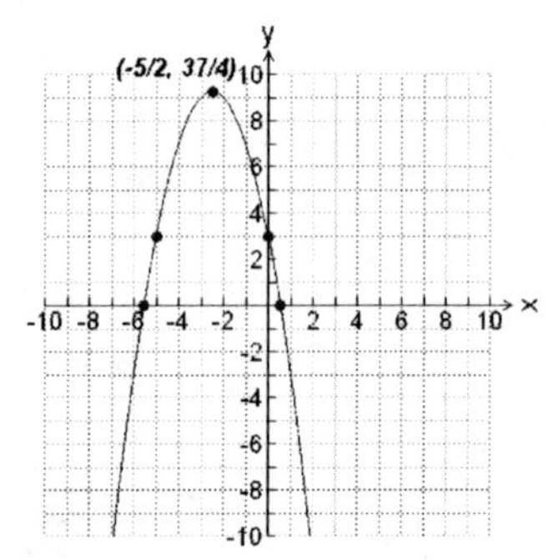

Objective 4

23. $y=(x-3)^2-4$; $(3,-4)$

25. $y=-(x-6)^2+2$; $(6,2)$

Objective 5

27. Vertex: $(1,\ -4)$, y-int: $(0,\ -3)$, x-int: $(-1,\ 0)$, $(3,\ 0)$

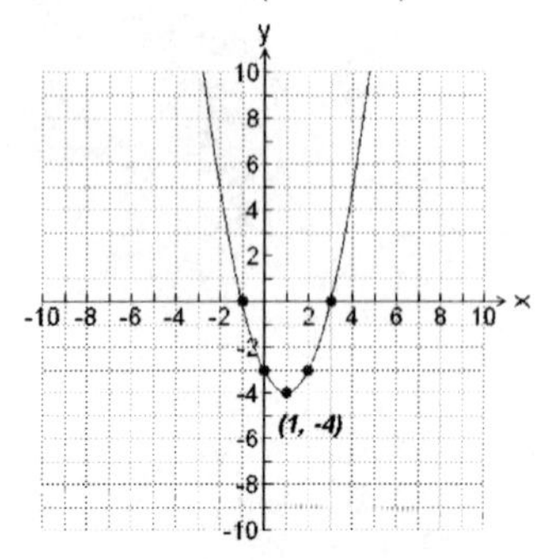

Objective 6

29. Vertex: $(4, \ -36)$, y-int: $(0, \ -20)$, x-int: $(-2, \ 0)$, $(10, \ 0)$

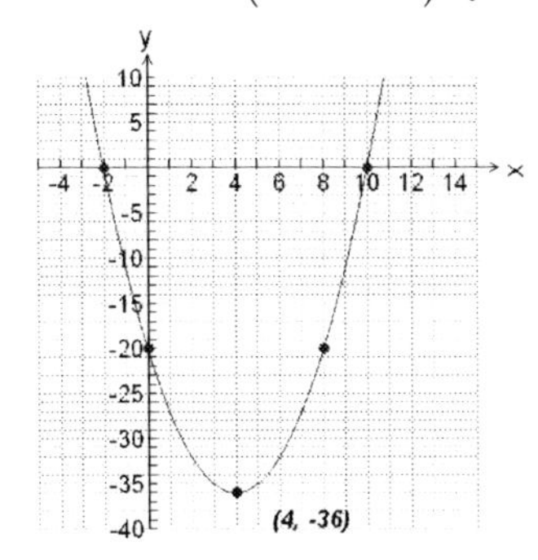

7.5 Applications Using Quadratic Equations

Objective 1

1. Base: 5 feet, Height: 13 feet
3. Base: 28 feet, Height: 5 feet
5. Base: 14 inches, Height: 6 inches
7. 3 inches by 7 inches
9. 40 feet by 40 feet

Objective 2

11. 23.2 feet
13. 19.1 feet

Objective 3

15. 2.9 feet
17. Length: 2.6 feet, Width: 1.6 feet
19. 16 feet

Objective 4

21. Minimum, 324
23. Maximum, $\frac{185}{4}$
25. Minimum, $\frac{199}{75}$
27. Minimum, 293

Objective 5

29. $20
31. $45
33. 100 ft by 100 ft, 10,000 sq ft

7.6 Quadratic and Rational Inequalities

Objective 1

1. $(-\infty, -4] \cup [2, \infty)$

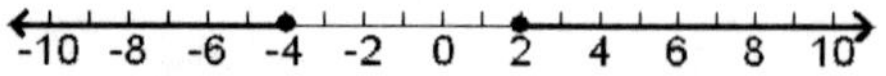

3. $[4-\sqrt{5}, 4+\sqrt{5}]$

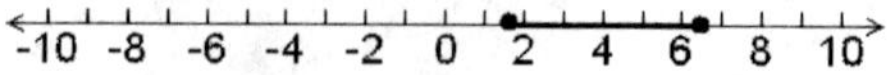

5. $[-6, -4]$

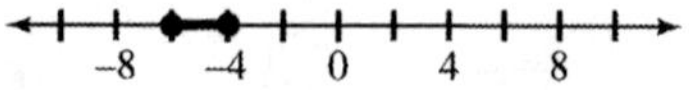

7. $\varnothing$

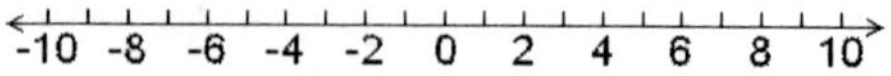

9. $(-\infty, -12] \cup [8, \infty)$

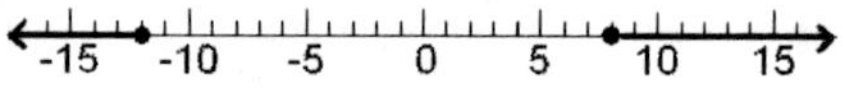

11. $(-\infty, -2] \cup [9, \infty)$

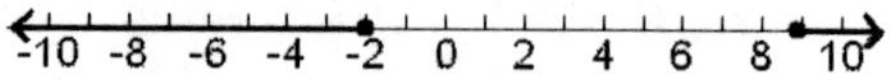

13. $[4-\sqrt{14}, 4+\sqrt{14}]$

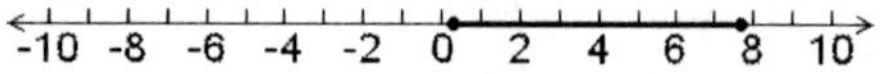

Objective 2

15. $(3, 8)$

Objective 3

17. $(-\infty, -2) \cup (0, 3) \cup (3, \infty)$

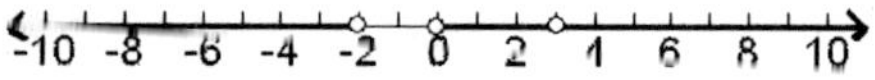

19. $(-\infty, -10) \cup (-6, 0) \cup (5, \infty)$

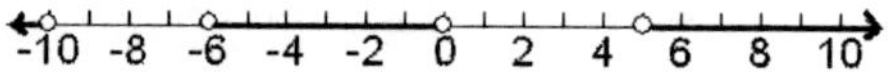

21. $(-\infty, -6] \cup (-4, -3) \cup [8, \infty)$

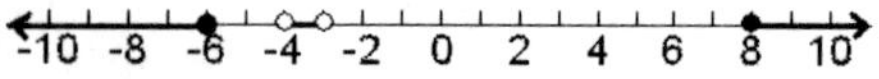

23. $(-9, 5)$

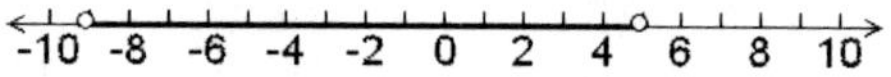

Objective 4

25. $\mathbb{R}$

27. $[-8, 6]$

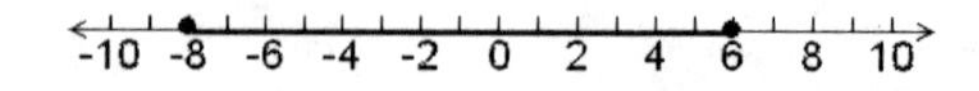

Objective 5

29. 2 sec to 5 sec

7.7 Other Functions and Their Graphs

Objective 1

1. Domain: $\left[\frac{1}{2},\ \infty\right)$, Range: $[5,\ \infty)$

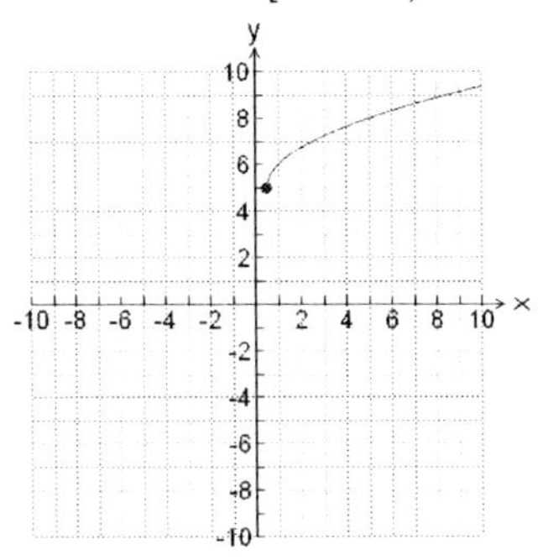

3. Domain: $(-\infty,\ 0]$, Range: $[0,\ \infty)$

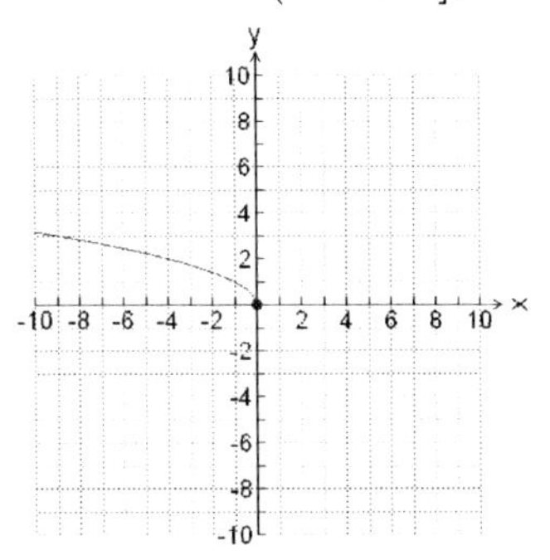

5. Domain: $[0,\ \infty)$, Range: $[4,\ \infty)$

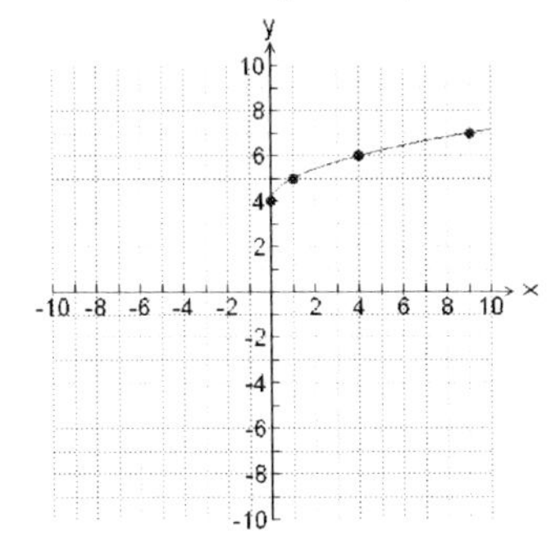

Objective 2

7. Domain: $(-\infty, \infty)$, Range: $(-\infty, \infty)$

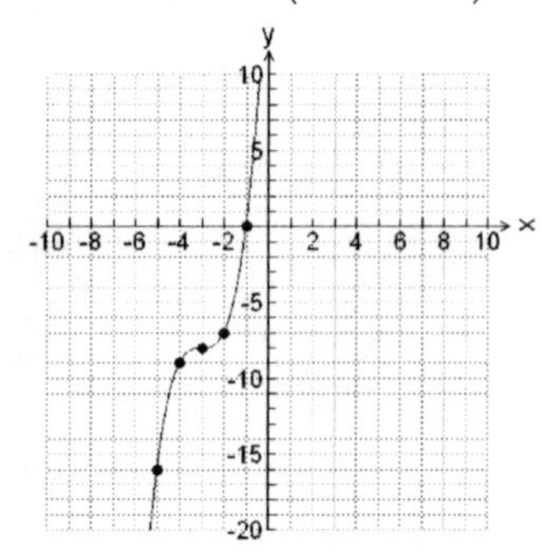

Objective 3

9. $f(x) = -\sqrt{x+5}$

11.a.

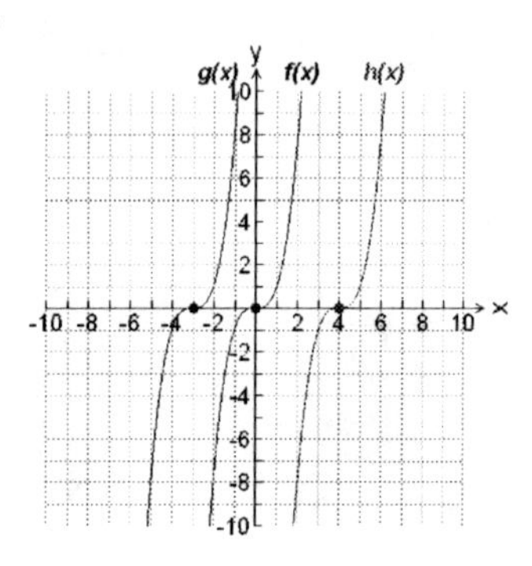

b. The graph of $f(x) = (x+22)^3$ would be 22 units to the left of $f(x) = x^3$, and the graph of $f(x) = (x-67)^3$ would be 67 units to the right of $f(x) = x^3$.

13.

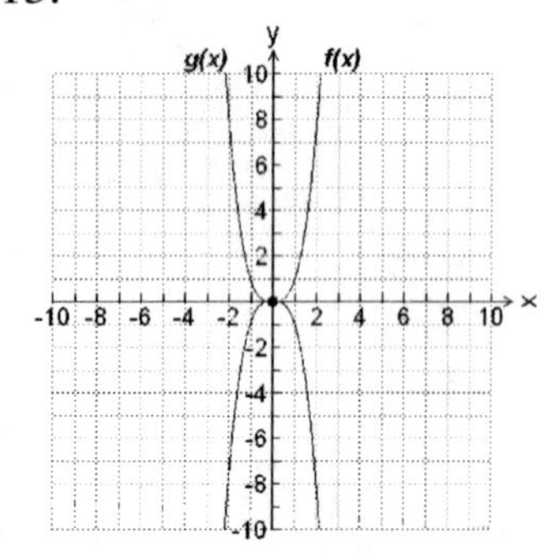

The graph of $g(x) = -x^3$ is the reflection of $f(x) = x^3$ about the *x*-axis.

Chapter 8 LOGARITHMIC AND EXPONENTIAL FUNCTIONS

8.1 The Algebra of Functions

Key Terms

1. $(f \circ g)(x)$; $(g \circ f)(x)$ 2. $(f+g)(x)$ 3. $(f \cdot g)(x)$

Objective 1

1. $x+6$ 3. $2x^2-16x-5$ 5. $\dfrac{1}{x+8}$ 7. -163

9. $-\dfrac{7}{5}$ 11. $g(x)=x-2$

Objective 2

13.

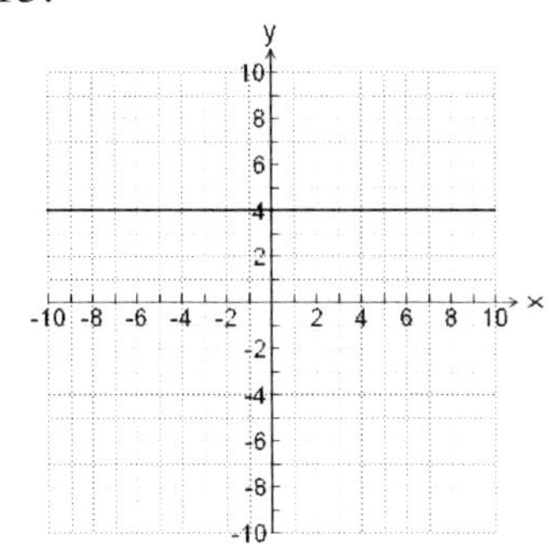

15.

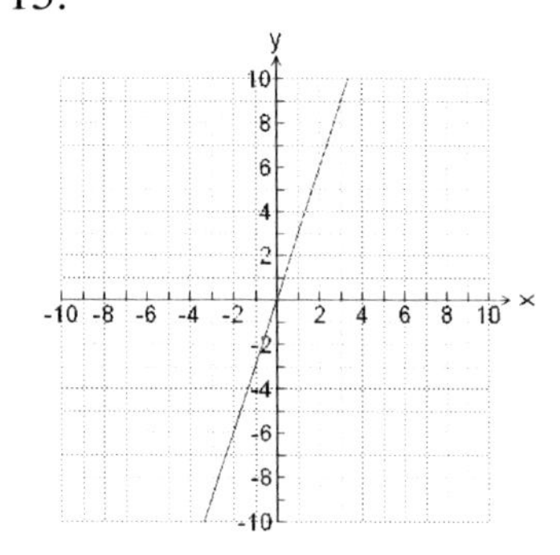

17.

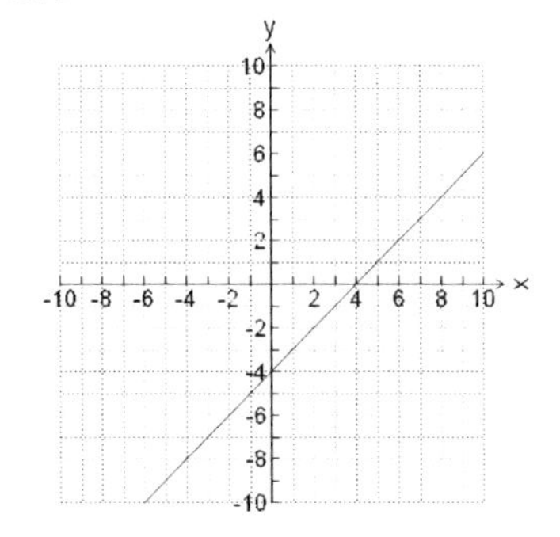

Objective 3

19.a. $(f-g)(x)=6x^2-97x+1978$, this is the U.S. population change (births – deaths), in thousands, x years after 1990   b. 7698, the U.S. population will grow by 7,698,000 in 2030

Objective 4

21. -40

Objective 5

23. $\sqrt{x^2-x-72}$, $(-\infty,\ -8]\cup[9,\ \infty)$ 25. $f(x)=\frac{x}{2}$

8.2 Inverse Functions

Key Terms

1. horizontal   2. domain   3. inverse

Objective 1

1. Yes   3. Yes   5. No; explanations will vary

7. No; explanations will vary

Objective 2

9. No

Objective 3, Objective 4

11. Not inverses   13. Inverses

Objective 5

15. $f^{-1}(x)=x$ 17. $f^{-1}(x)=\frac{7x+6}{2x}$ 19. $f^{-1}(x)=\sqrt{x}$, $x\geq 0$

21. $\{(7,\ 4),\ (8,\ 5),\ (9,\ 7),\ (10,\ 11),\ (11,\ 19)\}$

Objective 6

23.

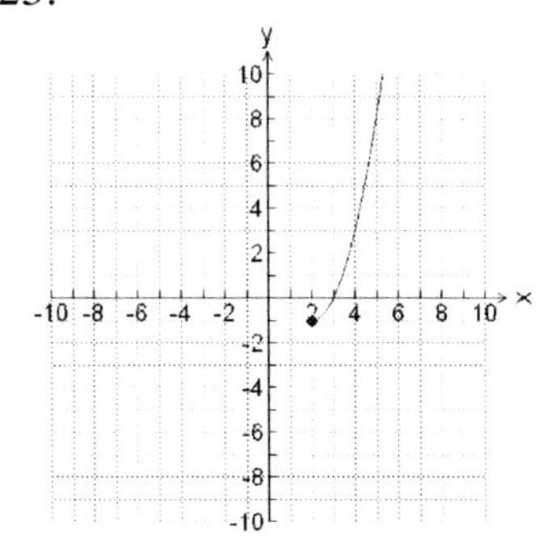

25.

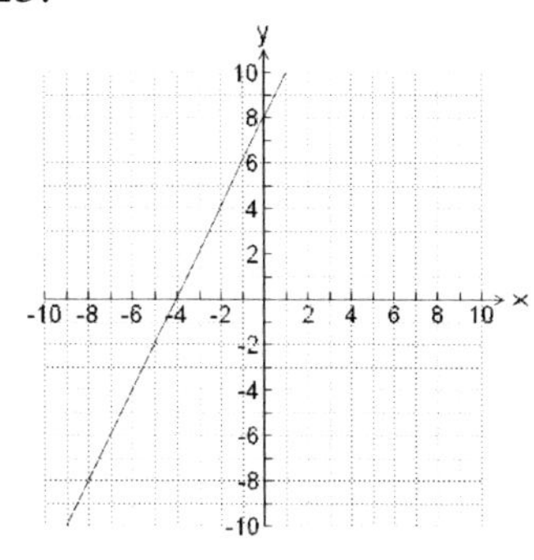

27.

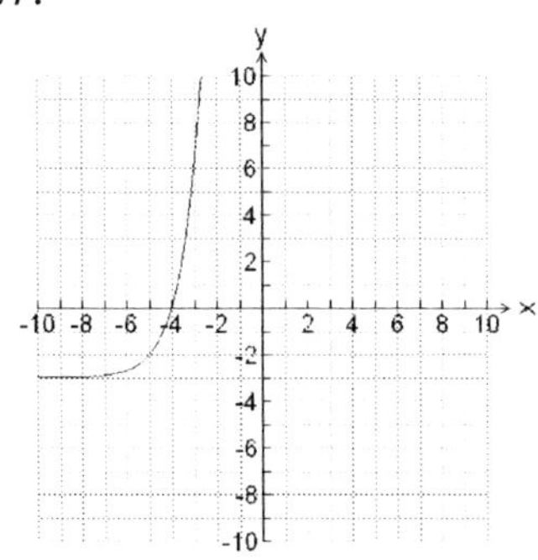

29.

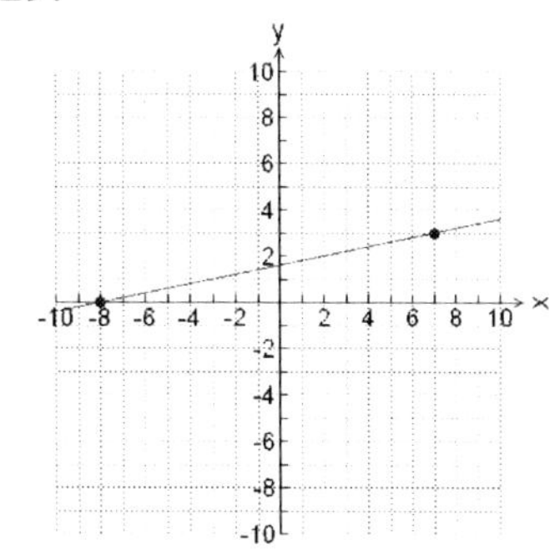

8.3 Exponential Functions

Key Terms

1. asymptote   2. base   3. increasing

Objective 1

1. $\frac{1}{8}$ 3. $\frac{5}{2}$

Objective 2

5. $-\frac{14}{9}$ 7. $\frac{313}{8}$ 9. $7\frac{1}{9}, 7\frac{1}{3}, 8, 10, 16$

Objective 3

11. Domain: $(-\infty, \infty)$, Range: $(-10, \infty)$

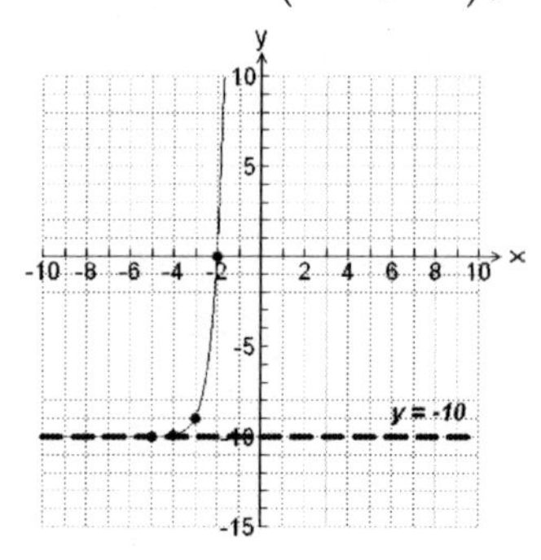

Objective 4

13. 1088.633

Objective 5

15. $\{3\}$ 17. $\{2\}$ 19. $\{-5\}$ 21. $\{-12, 2\}$

Objective 6

23. $602,257.52 25. 18,339 cells 27. 330 million 29. No

8.4 Logarithmic Functions

Key Terms

1. common 2. logarithmic; argument

Objective 1, Objective 2

1. 5 3. -5 5. 8 7.a. 2 b. -1

Objective 3

9. 2.538 11. 1.504 13. 0.693 15. 1.133

Objective 4

17. $\log_3 531,441 = 12$ 19. $\log_3 14 = x$ 21. $2^x = 22$ 23. $0.3^x = 307$

Objective 5

25. $\{9.518\}$ 27. $\{0.011\}$ 29. $\left\{\frac{1}{32}\right\}$

Objective 6

31.

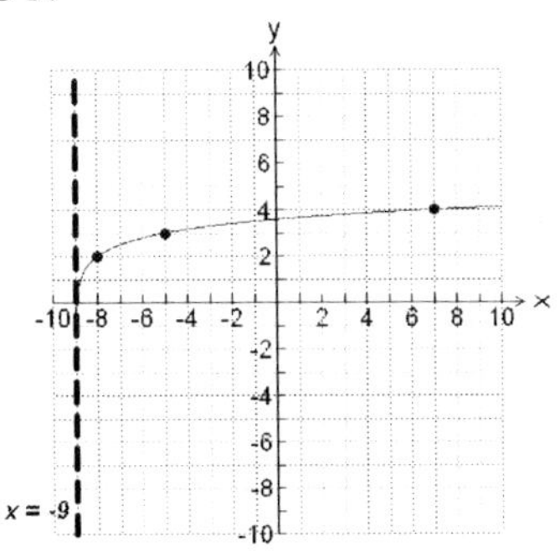

Objective 7

33. 84.7% 35. 1,584,893 times larger

8.5 Properties of Logarithms

Objective 1

1. $\log_c(FY)$ 3. $\log_8 3 + \log_8 w$

Objective 2

5. $\log_6 7$ 7. $\log_5 14 + \log_5 x - \log_5 y - \log_5 z$

Objective 3

9. $14\log_9 s$ 11. $\log_6 m^2$

Objective 4

13. 13 15. x

Objective 5

17. $\log_2 820{,}125$ 19. $\log_4\left(\frac{x^2-10x+21}{512}\right)$ 21. $\log_3\left(\frac{1}{30}\right)$

23. $\log_5\left(\frac{1}{136}\right)$ 25. $\log_6 36 = 2$

Objective 6

27. $\ln a + \ln b - \ln c$ 29. $6 + 12\log_3 a$

Objective 7

31. $D - B$ 33. $B + 2C$ 35. $15 + 15\log_2 x + 20\log_2 y$

37. -0.131 39. 0.862

8.6 Exponential and Logarithmic Equations

Objective 1

1. $\{-7\}$

Objective 2

3. $\{0.798\}$ 5. $\{-1.663\}$

Objective 3

7. $\{-2\}$ 9. $\{13\}$

Objective 4

11. $\left\{\frac{109}{2}\right\}$ 13. $\{0.097\}$ 15. $\{122\}$

Objective 5

17. $\{30\}$ 19. $\{13\}$ 21. $\{5.193\}$

Objective 6

23. $f^{-1}(x) = \ln(x+2) - 9$ 25. $f^{-1}(x) = 10^{x+8} - 9$

8.7 Applications of Exponential and Logarithmic Functions

Objective 1

1. 9.6% annual, 3.7 yr 3. 10.0 yr

Objective 2

5. 2002

Objective 3

7. 8.5 million tons

Objective 4

9. \$5496.26 11. In 2023 (47.6 yr)

Objective 5

13. 6, acid 15. 3.98×10^{-8} moles per liter 17. $10^{-0.3}$ watts per square meter

19. 74.4%

8.8 Graphing Exponential and Logarithmic Functions

Key Terms

1. all real numbers 2. logarithmic 3. exponential

Objective 1, Objective 2

1. x-int: $(-4.999,\ 0)$, y-int: $(0,\ 3.7)$ 3. x-int: None, y-int: $(0,\ 34)$

5. Asymptote: $y = 16$, Domain: $(-\infty,\ \infty)$, Range: $(16,\ \infty)$

7. Asymptote: $x = -7$, Domain: $(-7,\ \infty)$, Range: $(-\infty,\ \infty)$

9. Domain: $(4, \infty)$; Range: $(-\infty, \infty)$

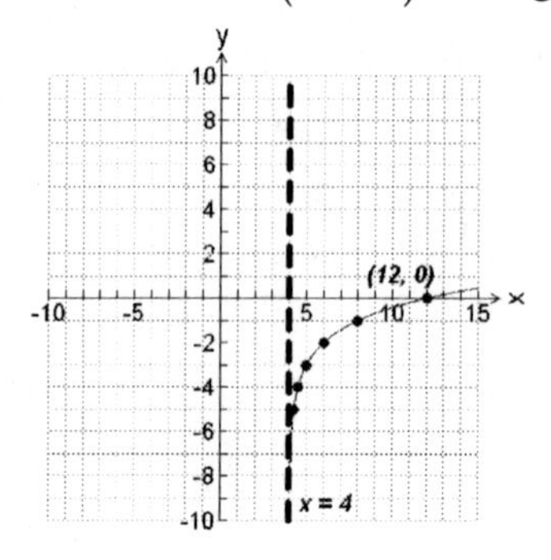

11. Domain: $(-\infty, \infty)$; Range: $(-12, \infty)$

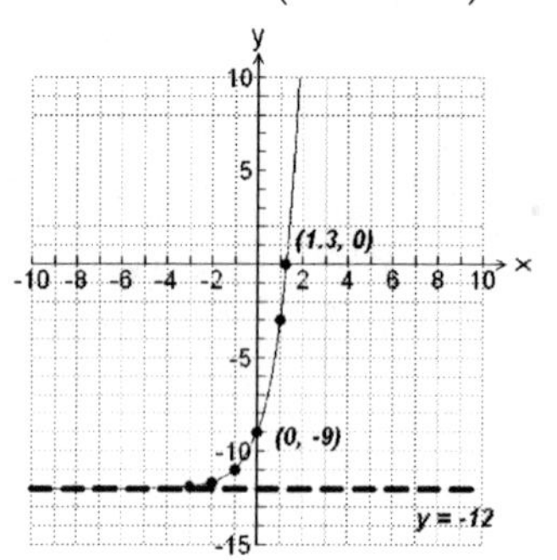

13. Domain: $(-8, \infty)$; Range: $(-\infty, \infty)$

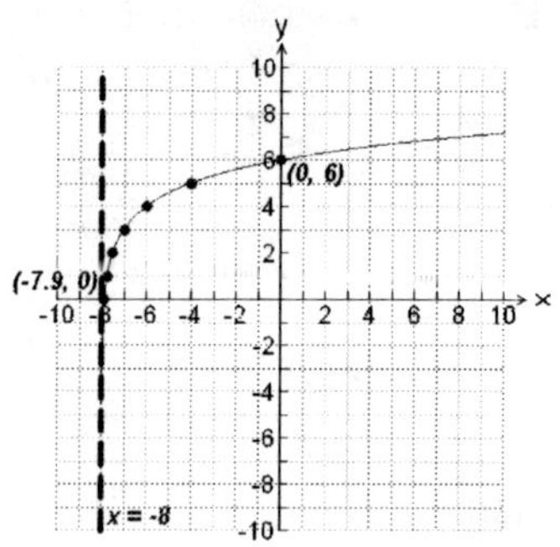

15. Domain: $(-\infty, \infty)$; Range: $(0, \infty)$

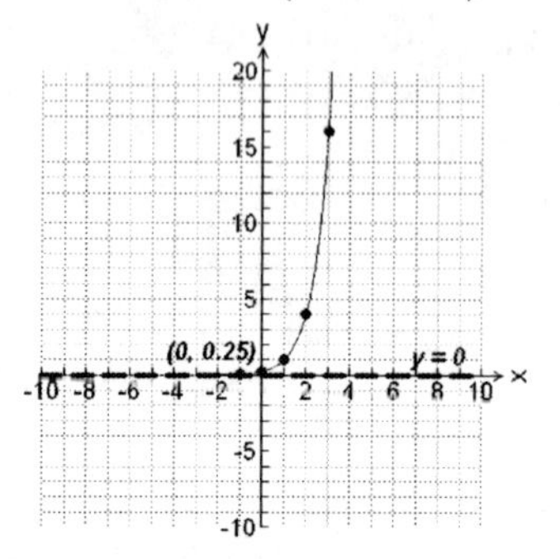

17. Domain: $(-\infty, \infty)$; Range: $(-12, \infty)$

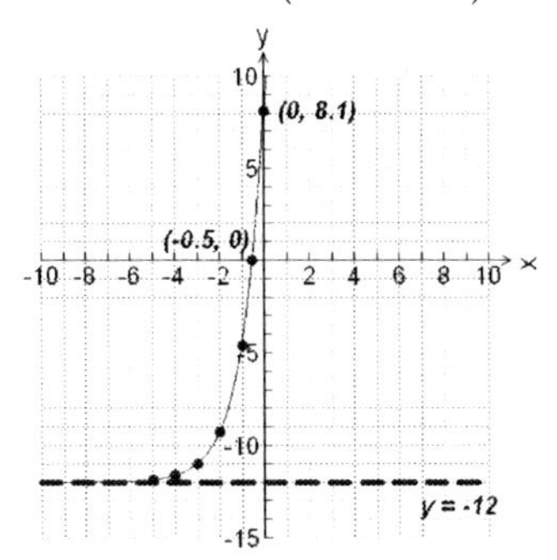

19. Domain: $(-4, \infty)$; Range: $(-\infty, \infty)$

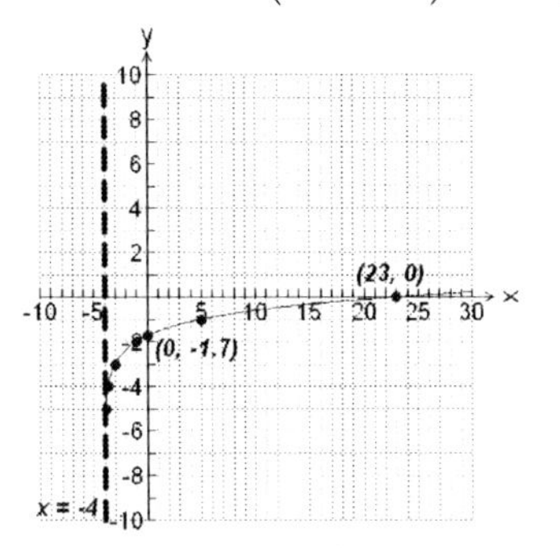

21. Domain: $(-\infty, \infty)$; Range: $(-9, \infty)$

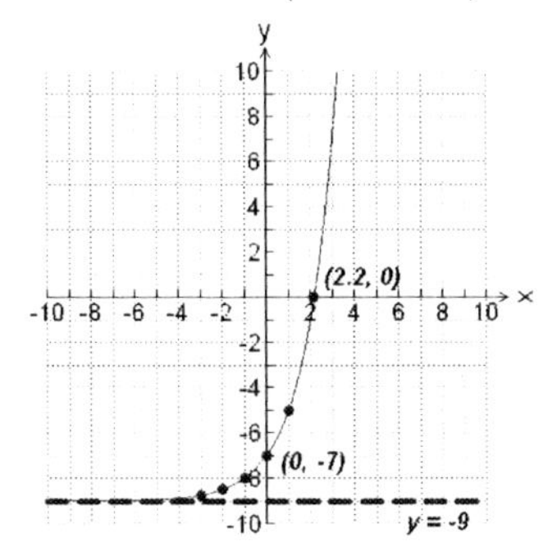

23.

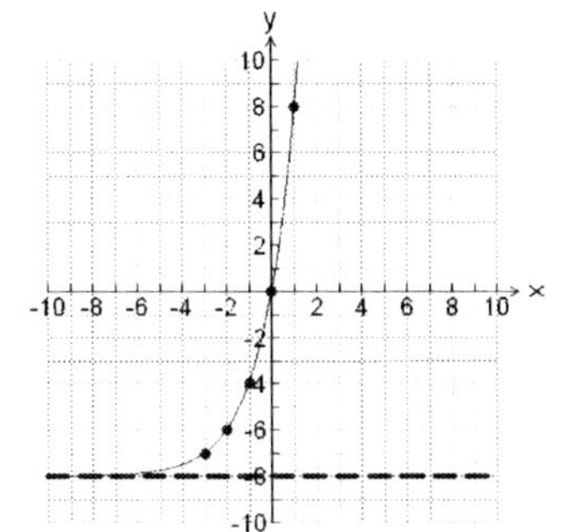

25. $f^{-1}(x) = e^{x+2} + 6$

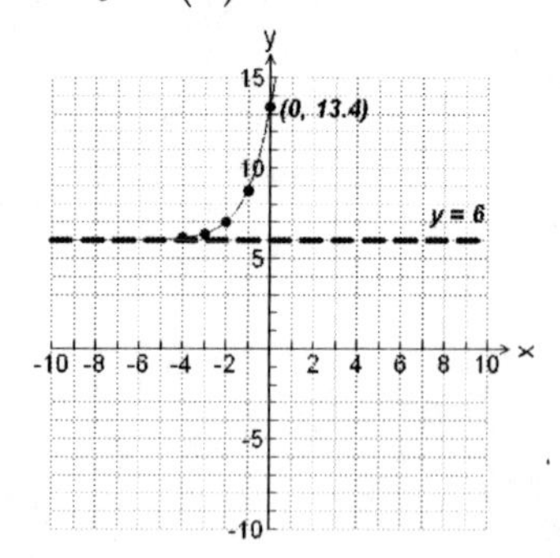

Chapter 9 CONIC SECTIONS

9.1 Parabolas

Key Terms

1. vertex 2. negative 3. positive

Objective 1

1. Vertex: $\left(-\frac{11}{2},\ -\frac{49}{4}\right)$; *y*-int: $(0,\ 18)$, *x*-int: $(-9,\ 0)$, $(-2,\ 0)$

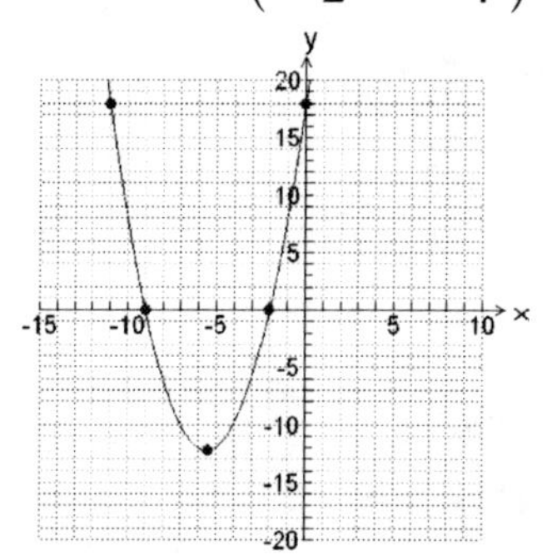

3. Vertex: $(-2,\ -8)$; *y*-int: $(0,\ -12)$, *x*-int: None

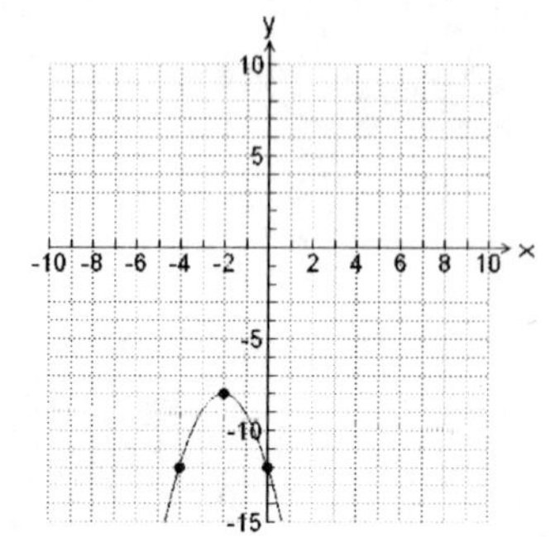

5. Vertex: $\left(\frac{7}{2}, \frac{101}{4}\right)$; y-int: $(0, 13)$, x-int: $\left(\frac{7 \pm \sqrt{101}}{2}, 0\right)$

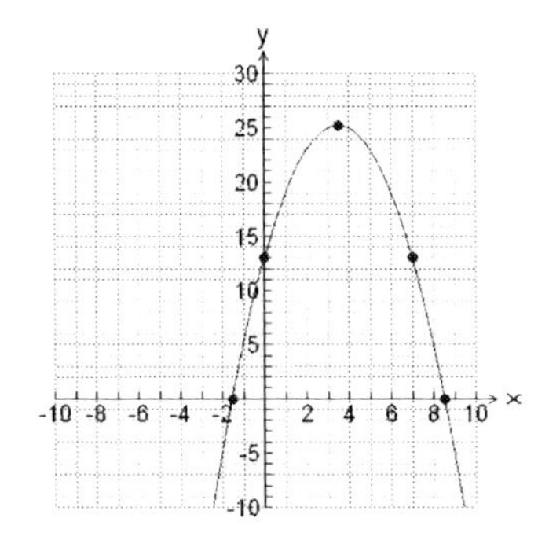

Objective 2

7. Vertex: $(5, -9)$; y-int: $(0, 16)$, x-int: $(2, 0)$, $(8, 0)$

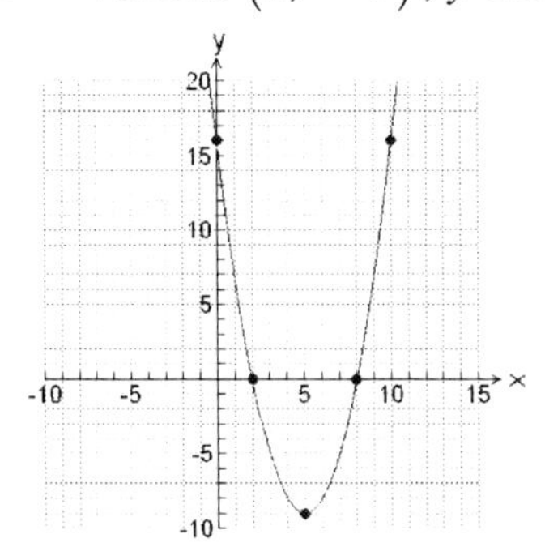

9. Vertex: $(-5, 3)$; y-int: $(0, 28)$, x-int: None

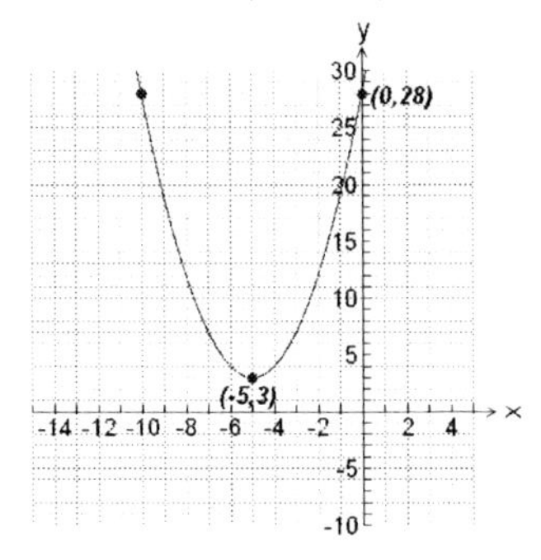

Objective 3

11. Vertex: $\left(-\dfrac{25}{4}, \dfrac{9}{2}\right)$; y-int: $(0, 2)$, $(0, 7)$, x-int: $(14, 0)$

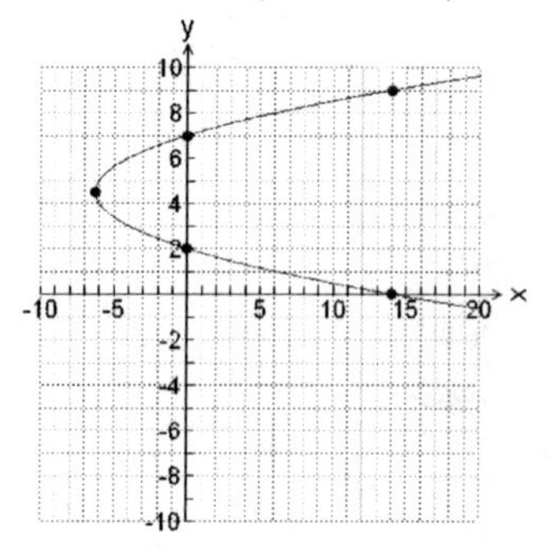

13. Vertex: $\left(-\dfrac{11}{4}, -\dfrac{7}{2}\right)$; y-int: None, x-int: $(-15, 0)$

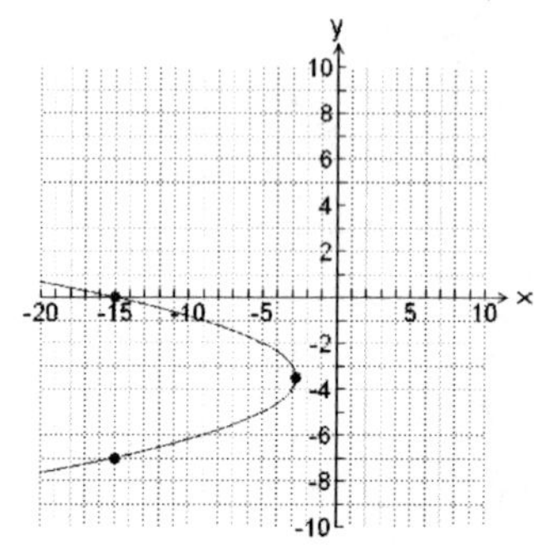

Objective 4

15. Vertex: $(-4, 3)$; y-int: $(0, 1)$, $(0, 5)$ x-int: $(5, 0)$

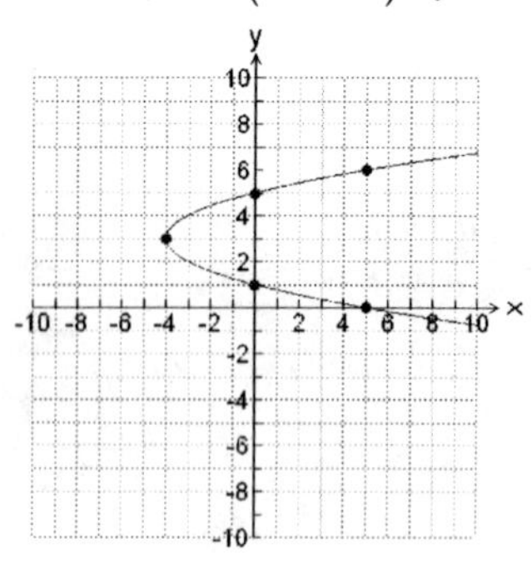

17. Vertex: $(9, \ -7)$; y-int: $(0, \ -10)$, $(0, \ -4)$ x-int: $(-40, \ 0)$

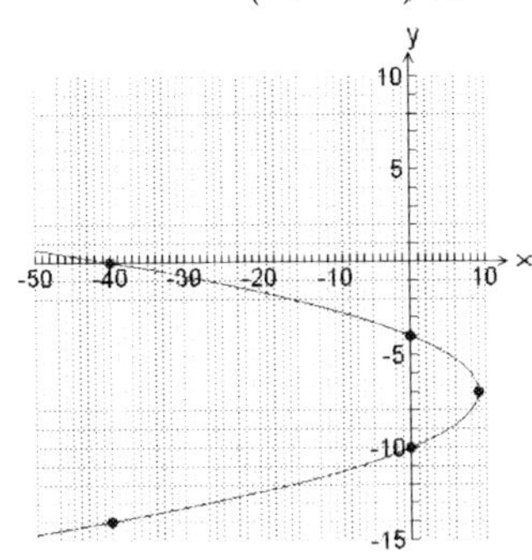

19. Vertex: $(6, \ -2)$; y-int: None, x-int: $(10, \ 0)$

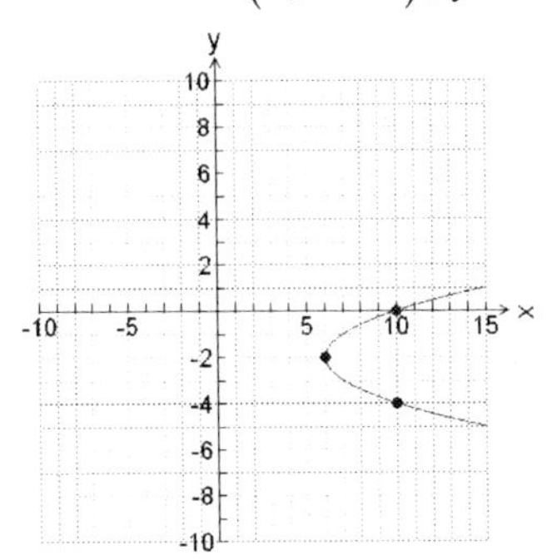

Objective 5

21. $\left(\frac{5}{2}, \ -\frac{205}{4}\right)$ 23. $\left(-\frac{77}{4}, \ -\frac{3}{2}\right)$

Objective 6

25. $y = -x^2 + 5x + 6$

9.2 Circles

Key Terms

1. distance 2. midpoint 3. center

Objective 1

1. 10.8 3. 10 5. $\left(3, \ -\frac{17}{2}\right)$ 7. $(5, \ 1)$

Objective 2

9. Center $(0,\ 0)$, $r = 6$

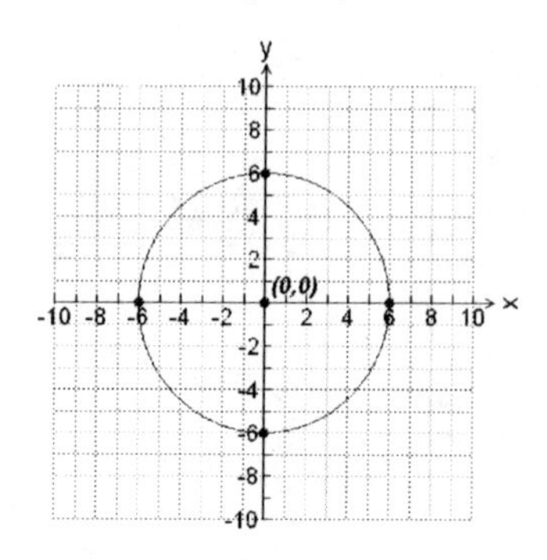

11. Center: $(0,\ 0)$, $r = \dfrac{5}{4}$

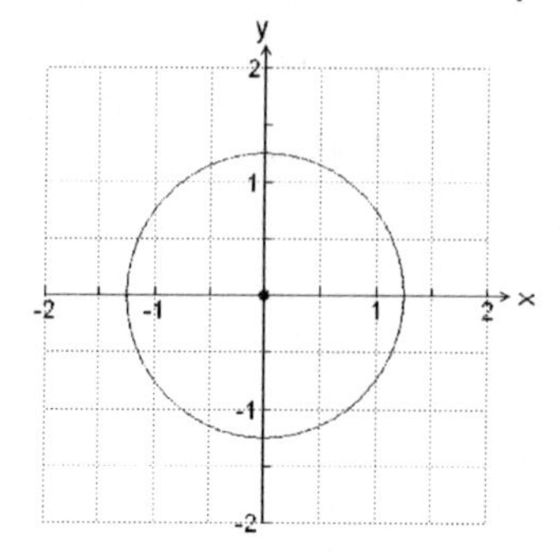

13. Center: $(0,\ 0)$, $r = 2\sqrt{6}$

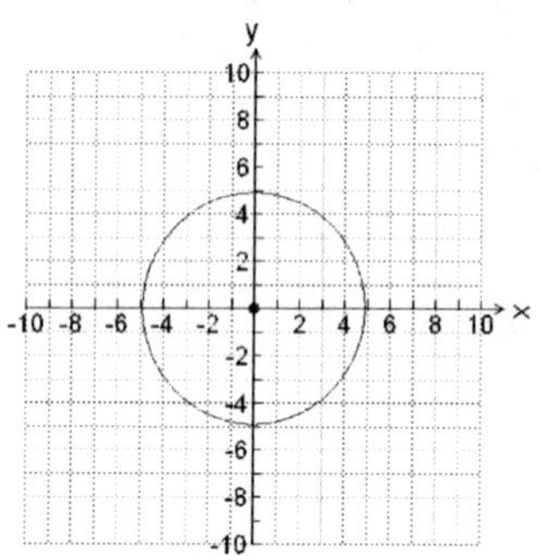

Objective 3

15. Center: $(-2,\ 7)$, $r = 4$

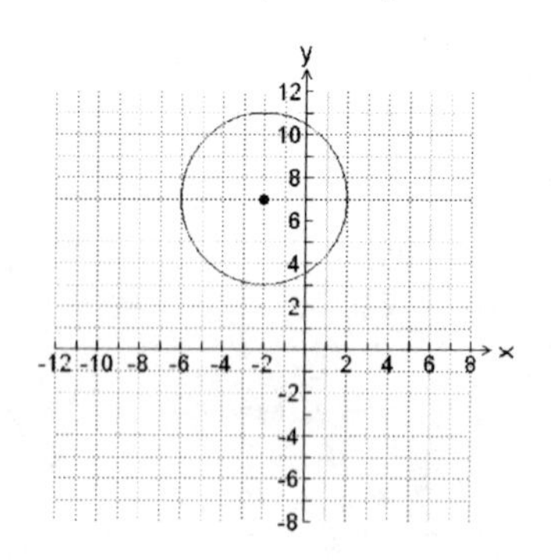

17. Center: $\left(\dfrac{17}{2},\ -\dfrac{1}{2}\right)$, $r = \dfrac{3}{2}$

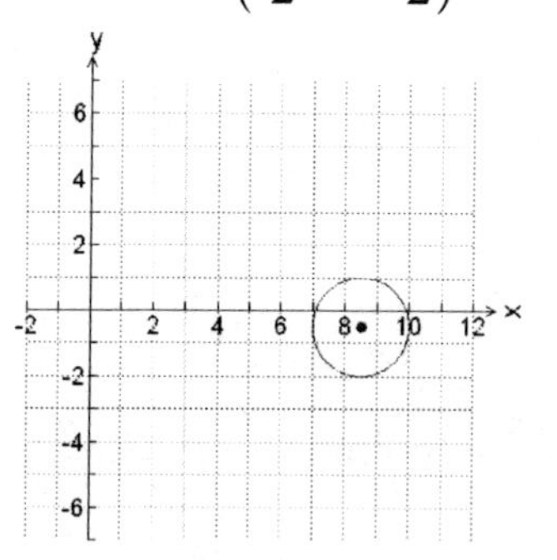

Objective 4

19. $(8,\ 10)$, 12

21. $\left(-\dfrac{5}{2},\ -\dfrac{7}{2}\right)$, 4

Objective 5

23. $x^2 + y^2 = 49$

25. $(x-3)^2 + (y-9)^2 = 25$

27. $\left(x - \frac{9}{2}\right)^2 + \left(y + \frac{7}{2}\right)^2 = \frac{1}{2}$

9.3 Ellipses

Key Terms

1. ellipse
2. minor
3. co-vertices

Objective 1

1. Center: $(0,\ 0)$, $a = 4$, $b = 9$

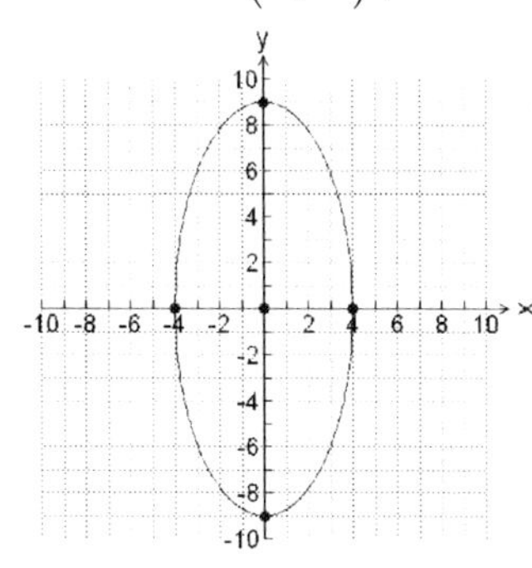

3. Center: $(0,\ 0)$, $a = 3\sqrt{2}$, $b = 7$

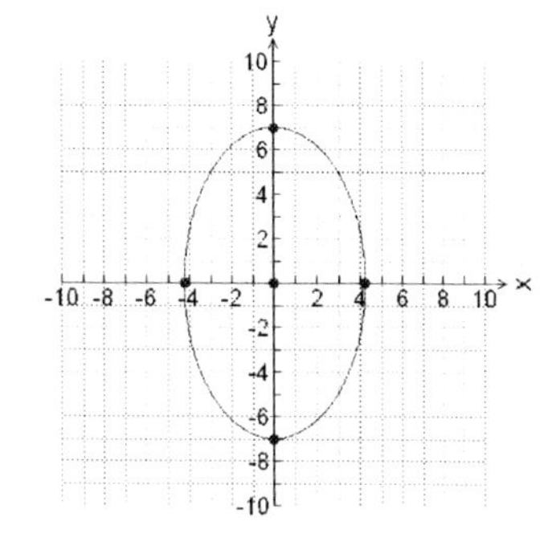

5. Center: $(0,\ 0)$, $a = 5$, $b = 2\sqrt{5}$

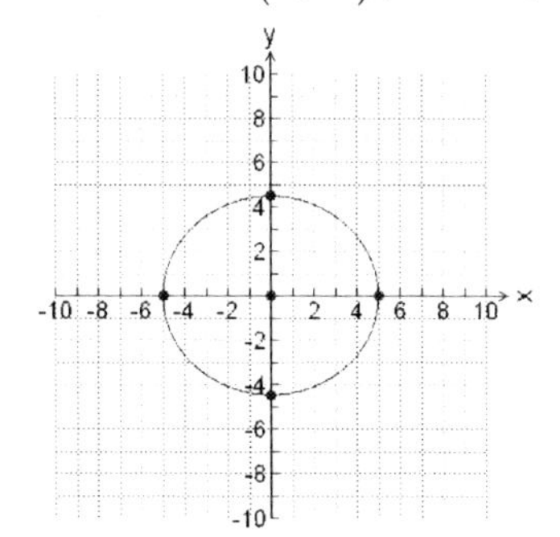

Objective 2

7. Center: $(3,\ -2)$, $a=8$, $b=4$

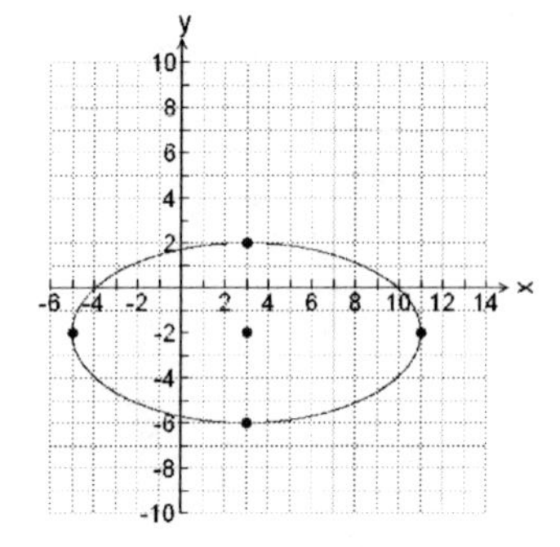

9. Center: $(-4,\ -9)$, $a=5$, $b=10$

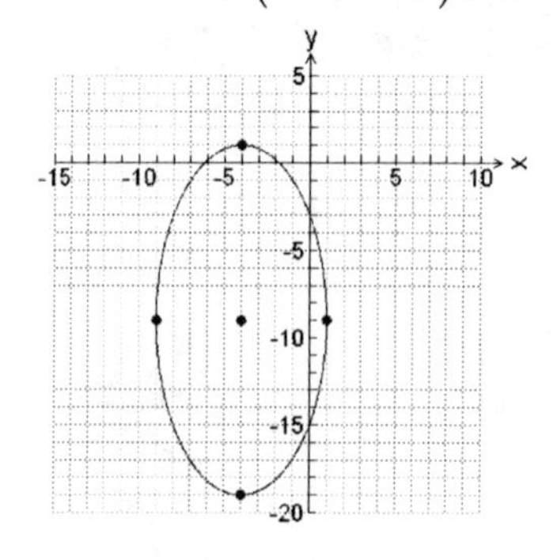

11. Center: $(-3,\ 4)$, $a=3$, $b=4\sqrt{2}$

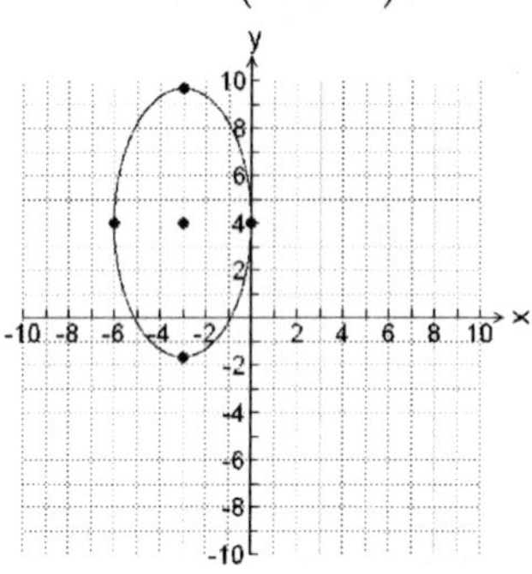

Objective 3

13. Center: $(5,\ -3)$, $a=10$, $b=6$

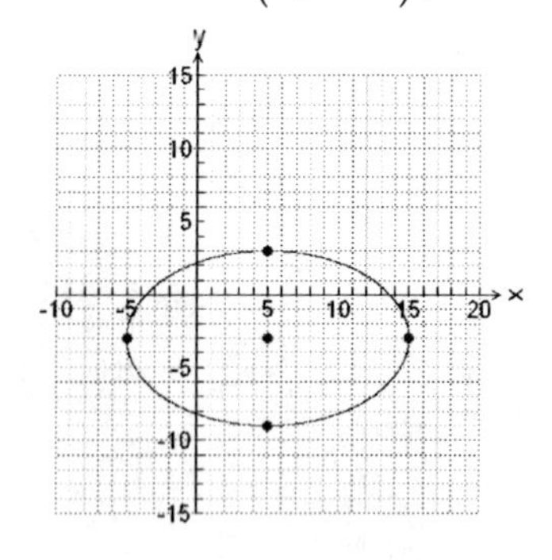

15. Center: $(3,\ -7)$, $a=2$, $b=10$

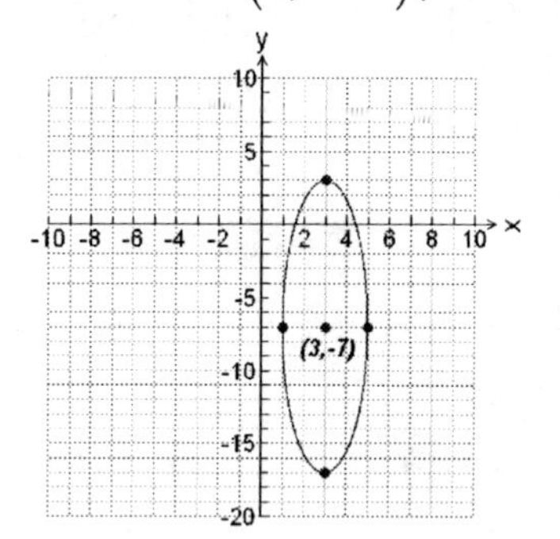

Objective 4

17. $\dfrac{(x-7)^2}{16}+\dfrac{(y-1)^2}{36}=1$

19. $\dfrac{(x+5)^2}{4}+\dfrac{(y+6)^2}{9}=1$

21. $\dfrac{(x-4)^2}{25}+(y+7)^2=1$

23. $\dfrac{x^2}{49}+\dfrac{(y-3)^2}{9}=1$

9.4 Hyperbolas

Key Terms

1. transverse
2. asymptote
3. foci
4. conjugate

Objective 1

1. Center: $(0,\ 0)$, $a=9$, $b=5$

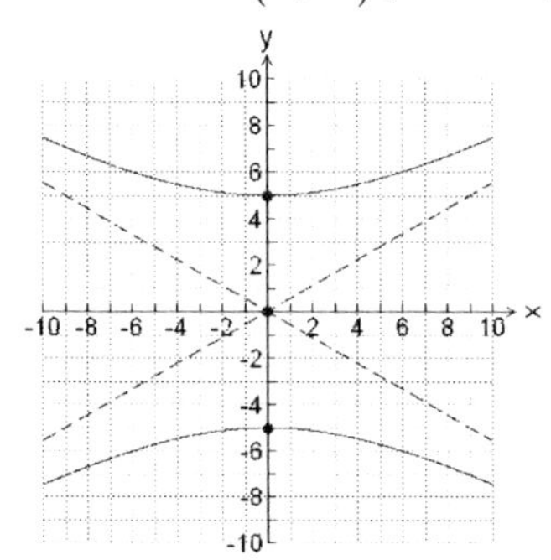

3. Center: $(0,\ 0)$, $a=2\sqrt{2}$, $b=5$

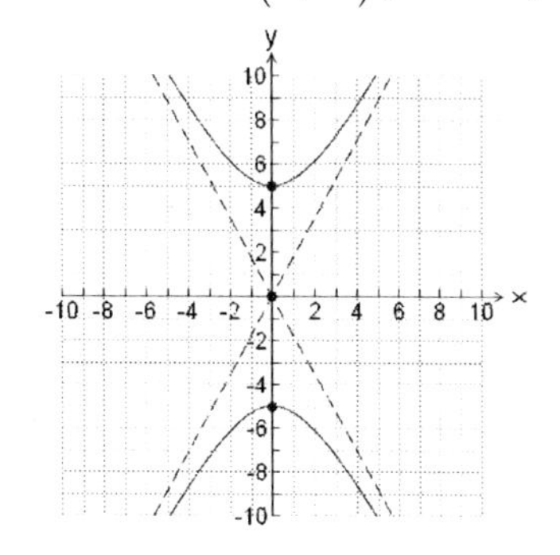

Objective 2

5. Center: $(-3,\ 4)$, $a=2$, $b=3$

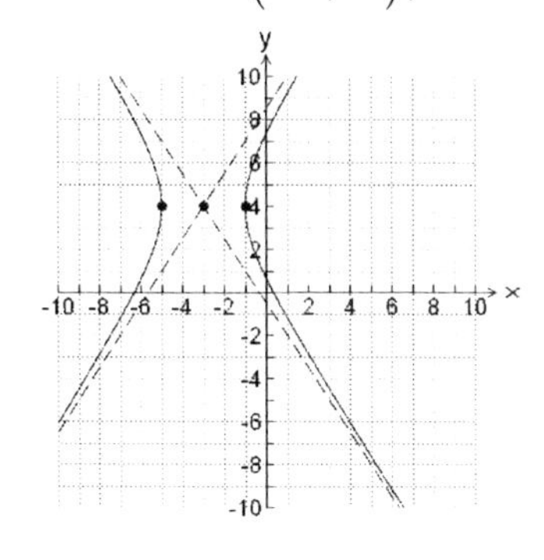

7. Center: $(-6,\ 3)$, $a=5$, $b=4$

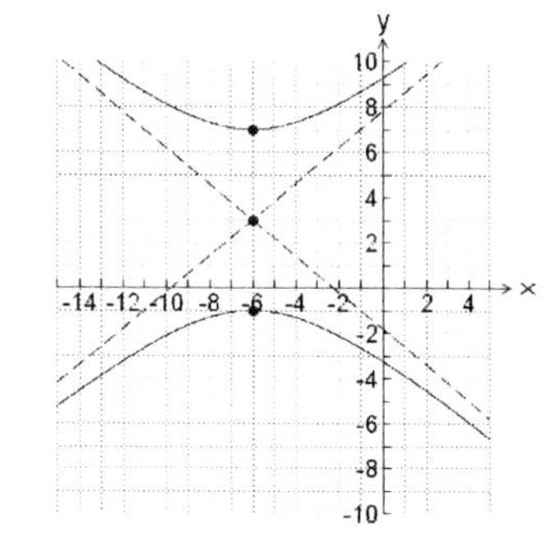

9. Center: $(8,\ 6)$, $a=\sqrt{3}$, $b=2\sqrt{2}$

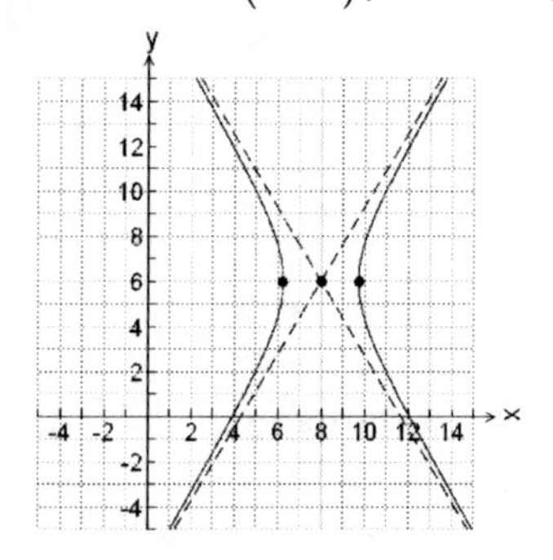

Objective 3

11. Center: $(-6,\ 0)$, $a=5$, $b=6$

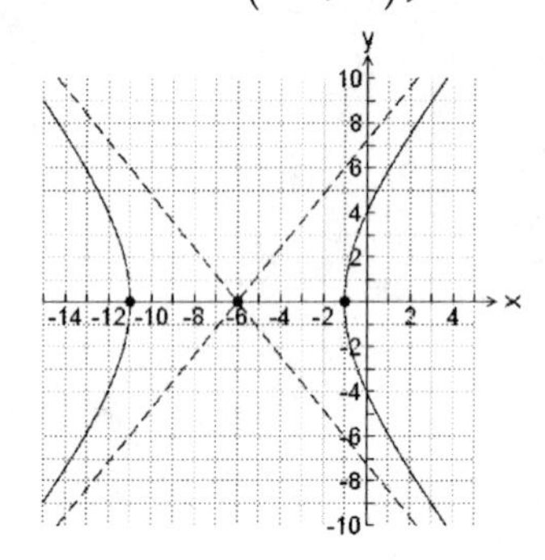

13. Center: $(4,\ -1)$, $a=4$, $b=3$

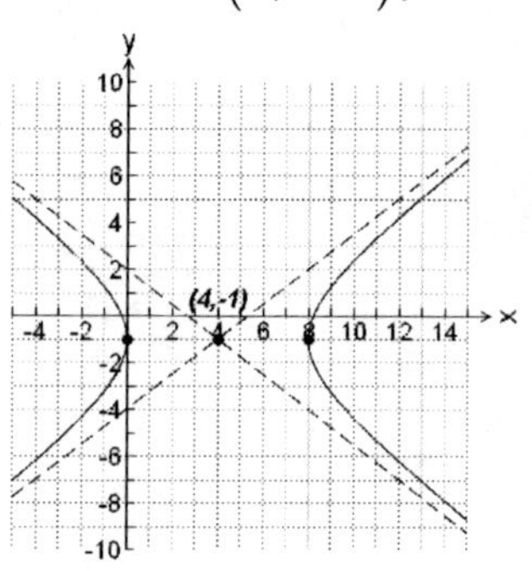

Objective 4

15. $\dfrac{(x-8)^2}{64}-\dfrac{(y+9)^2}{81}=1$

17. $\dfrac{x^2}{4}-\dfrac{y^2}{49}=1$

19. $\dfrac{x^2}{25}-\dfrac{(y+2)^2}{4}=1$

9.5 Nonlinear Systems of Equations

Objective 1

1. $(7,\ 4),\ (-8,\ -1)$

3. $(3,\ -2)$

5. $(10,\ 3),\ (-26,\ -1)$

7. $(-3,\ 4)$

9. $(35,\ 27),\ (5,\ -3)$

11. $(8,\ -6),\ (-8,\ -6)$

13. $(-4,\ 0),\ \left(\dfrac{20}{9},\ \dfrac{2\sqrt{14}}{3}\right),\ \left(\dfrac{20}{9},\ -\dfrac{2\sqrt{14}}{3}\right)$

15. $(6,\ 61),\ (10,\ 289)$

Objective 2

17. $\left(3\sqrt{3},\ 6\right), \left(3\sqrt{3},\ -6\right), \left(-3\sqrt{3},\ 6\right), \left(-3\sqrt{3},\ -6\right)$

19. $\left(3,\ \sqrt{3}\right), \left(3,\ -\sqrt{3}\right), \left(-3,\ \sqrt{3}\right), \left(-3,\ -\sqrt{3}\right)$

21. $\left(\frac{\sqrt{10}}{4},\ \frac{3\sqrt{6}}{4}\right), \left(\frac{\sqrt{10}}{4},\ -\frac{3\sqrt{6}}{4}\right), \left(-\frac{\sqrt{10}}{4},\ \frac{3\sqrt{6}}{4}\right), \left(-\frac{\sqrt{10}}{4},\ -\frac{3\sqrt{6}}{4}\right)$

23. $\left(\sqrt{2},\ 5\right), \left(\sqrt{2},\ -5\right), \left(-\sqrt{2},\ 5\right), \left(-\sqrt{2},\ -5\right)$

Objective 3

25. 7 in. by 24 in. 27. 24 ft by 40 ft 29. 11 in. by 60 in.

Chapter 10 SEQUENCES, SERIES, AND THE BINOMIAL THEOREM

10.1 Sequences and Series

Key Terms

1. alternating 2. finite 3. lower 4. series

Objective 1

1. 55 3. $\frac{2}{3}, \frac{4}{9}, \frac{8}{27}, \frac{16}{81}, \frac{32}{243}$ 5. $-4,\ -8,\ -12,\ -16,\ -20$

Objective 2

7. $1024,\ -4096,\ 16{,}384,\ a_n = (-1)^n \cdot 4^{n-1}$

9. $-64,\ 81,\ -100,\ a_n = (-1)^{n+1}(n+2)^2$ 11. $\frac{1}{2}, \frac{7}{4}, \frac{11}{2}, \frac{67}{4}, \frac{101}{2}$

13. $0.6,\ -0.7,\ 1.9,\ -3.3,\ 7.1$

Objective 3

15. 31.875 17. 112 19. 4094 21. $-314,574$

Objective 4

23. 115 25. -12 27. $\sum_{i=1}^{5}(13-5i)$ 29. $\sum_{i=1}^{4}(-1)^{i+1}\cdot i^3$

Objective 5

31. a. 200, 225, 250, 275, 300, 325, 350 b. 1925

10.2 Arithmetic Sequences and Series

Key Terms

1. general 2. common 3. sequence

Objective 1

1. -4 3. -19 5. $\frac{3}{4}$ 7. $\frac{5}{12}$ 9. -5

11. 7, 20, 33, 46, 59 13. 78, 117, 156, 195, 234

15. $\frac{3}{8}, \frac{51}{8}, \frac{99}{8}, \frac{147}{8}, \frac{195}{8}$ 17. $a_n = -1.1n + 5.4$

19. -193 21. $\frac{567}{4}$ 23. 211 25. 222nd

Objective 2

27. 16,330 29. $-15,264$ 31. 15,625

Objective 3

33. a. $a_n = n + 7$ b. 15

35. a. \$7500, \$8500, \$9500, \$10,500, \$11,500, \$12,500, \$13,500, \$14,500

b. $a_n = 1000n + 6500$ c. \$456,000

10.3 Geometric Sequences and Series

Key Terms

1. general 2. ratio 3. partial

Objective 1

1. $\frac{1}{4}$ 3. $-\frac{4}{3}$ 5. $\frac{1}{32}, -\frac{1}{8}, \frac{1}{2}, -2, 8$

7. 0.2, 0.02, 0.002, 0.0002, 0.00002

Objective 2

9. $a_n = 54 \cdot \left(\frac{1}{3}\right)^{n-1}$ 11. 43,046,721 13. $\frac{3}{16,384}$ 15. $-\frac{128}{3125}$

Objective 3

17. 18.25824 19. 20,478.75

Objective 4

21. Yes, 30 23. Yes, $0.\overline{18}$ or $\frac{2}{11}$ 25. Yes, 50.4 or $\frac{252}{5}$

Objective 5

27. \$18,000, \$10,800, \$6480, \$3888

10.4 The Binomial Theorem

Key Terms

1. factorial 2. binomial

Objective 1

1. 39,916,800 3. 479,001,600 5. 6 7. 123! 9. 13

11. 10 13. 362,874 15. 1287

Objective 2

17. 56　　19. 9

Objective 3

21. $x^6 - 6x^5y + 15x^4y^2 - 20x^3y^3 + 15x^2y^4 - 6xy^5 + y^6$

23. $25x^2 + 20xy + 4y^2$

25. $2{,}097{,}152x^7 - 12{,}845{,}056x^6y + 33{,}718{,}272x^5y^2 - 49{,}172{,}480x^4y^3$
$+43{,}025{,}920x^3y^4 - 22{,}588{,}608x^2y^5 + 6{,}588{,}344xy^6 - 823{,}543y^7$

27. $x^5 - 5x^4y + 10x^3y^2 - 10x^2y^3 + 5xy^4 - y^5$

29. $x^5 - 25x^4y + 250x^3y^2 - 1250x^2y^3 + 3125xy^4 - 3125y^5$

Objective 4

31. $81x^4 + 216x^3y + 216x^2y^2 + 96xy^3 + 16y^4$

33. $7776x^5 - 6480x^4y + 2160x^3y^2 - 360x^2y^3 + 30xy^4 - y^5$

35. $32x^5 - 80x^4y + 80x^3y^2 - 40x^2y^3 + 10xy^4 - y^5$

Appendix A　SYNTHETIC DIVISION

1. $x-2$　　3. $x-8$　　5. $3x+8+\dfrac{21}{x-3}$　　7. $3x+14+\dfrac{65}{x-5}$

9. $2x^2 - 6x - 1 - \dfrac{5}{x+3}$